Sadguru Model of Rural Development

Pioneered by India's Social Work Stalwarts

The Author

Govindasamy Agoramoorthy is Distinguished Research Professor at the College of Environment and Health Sciences, Tajen University, Taiwan. His research ranges from nature conservation to sustainable development, and he has carried out field research related to nature conservation in Asia, Africa and South America during the last 25 years. He was Visiting Scientist at Smithsonian Institution in Washington DC (USA) from 1989 to 1993, and later served as an Adjunct Professor at the Ohio State University (Columbus, USA). His recent work targets multidisciplinary research on environment and development in Asia. He currently serves as Tata Visiting Chair in India at Sadguru Foundation (Dahod) and Research Advisor at SVYAS Yoga University (Bangalore). Professor Agoramoorthy has authored over 25 books, 100 book chapters and 500 scientific articles in peer-reviewed journals.

Sadguru Model of Rural Development

Pioneered by India's Social Work Stalwarts

Govindasamy Agoramoorthy

2017

Daya Publishing House®

A Division of

Astral International Pvt. Ltd.

New Delhi – 110 002

Publisher's Note:

Cataloging in Publication Data--DK
Courtesy: D.K. Agencies (P) Ltd. <docinfo@dkagencies.com>

Agoramoorthy, Govindasamy, 1957- author.
Sadguru model of rural development : pioneered by India's social work stalwarts / Govindasamy Agoramoorthy.
pages cm
Includes bibliographical references.
ISBN 9789387057777 (Int. Edition)

1. Sadguru (Organization : India). 2. Rural development--India--Rajasthan. 3. Dams--Environmental aspects--India--Rajasthan. 4. Irrigation farming--India--Rajasthan. 5. Arid regions agriculture--India--Rajasthan. I. Title. II. Title: Pioneered by India's social work stalwarts.

HN690.Z9C6 2017   DDC 307.141209544   23

Published by   :   **Daya Publishing House®**
A Division of
Astral International Pvt. Ltd.
– ISO 9001:2015 Certified Company –
4736/23, Ansari Road, Darya Ganj
New Delhi-110 002
Ph. 011-43549197, 23278134
E-mail: info@astralint.com
Website: www.astralint.com

— Dedicated to —

India's Legendary Rural Development Stalwarts,
Shri Harnath Jagawat and
Smt Sharmishta Jagawat

For sharing their immense knowledge and compassion

The support of Sir Dorabji Tata Trust (Mumbai) to conduct research at
Sadguru Foundation through the Tata-Sadguru Visiting Chair
Status awarded to the author is greatly appreciated!

Profound gratitude to

The Honorable Chief Minister of Rajasthan, Smt. Vasundhara Raje

For her gracious support to promote eco-friendly check dams in Rajasthan

Acknowledgments

"Gratitude is the fairest blossom which springs from the soul"— *HW Beecher*

I sincerely thank Shri Harnath Jagawat and Smt. Sharmishtha Jagawat for their kind hospitality and support during my numerous visits to Sadguru Foundation's project sites. I am fortunate to work with India's prominent rural development stalwarts for over a decade. Their vision and willingness to share their unique rural development social work experience have resulted in several high quality research papers published in high impact journals ultimately benefiting the international scientific community.

I am grateful to my colleagues at Sadguru Foundation namely Kanhaiya Choudhary, Sunita Chaudhary, Hitesh Shah, Ramesh Patel and Shodhan Shah for sharing their knowledge on developmental issues that range from lift irrigation to check dams, and from farming practices to financial management. Several staff members and volunteers affiliated to Sadguru Foundation interacted with me during numerous field visits in Gujarat, Rajasthan and Madhya Pradesh states. They answered all my queries on rural development without any hesitation, so I thank them all for their kind cooperation. During each village visit, I had the unique opportunity to encounter large number of good-hearted farming folks who assisted me in many ways, from offering food to sharing their knowledge on the beauty of living in the wilderness.

Several youngsters from Chosala village tirelessly worked at the Sadguru Foundation's kitchen and I am grateful to the lead chef, Naginbhai, who took special care of me over the years by preparing delicious less-spicy food on a timely fashion. I thank them all for their compassion, care and diligence. I thank all the senior employees of Sadguru Foundation who tolerated my exploration into their professional activities with elegance, tolerance and candidness. Finally, I thank the financial support of Sir Dorabji Tata Trust, Mumbai for awarding me the Tata Visiting Chair status to carry out long-term scientific research at Sadguru Foundation.

Author and friends from
Sadguru Foundation, India

Preface

"Civilization as it is known today could not have evolved, nor can it survive, without an adequate food supply"—N. Borlaug

India is the seventh largest country in the world covering an area of 3.2 million square kilometers. Huge rivers namely the Ganges and Brahmaputra in east, Indus in west, Narmada in central and Cauvery in south run wild, and all these major rivers are considered sacred in Hinduism. Ancient rock shelters and cave paintings at Bhimbetka in Madhya Pradesh state are the earliest records of human life dating back to 10,000 BC. India is considered as the cradle of humanity since the Indus Valley Civilization originated there representing the advanced human civilization in 2,500 BC.

I am always thrilled to visit remote parts of India, especially the drylands since the country is dominated by villages. During my visits to the tribal drylands of Rajasthan, Madhya Pradesh and Gujarat states, I have seen several hamlets without electricity, infrastructure, sanitation and other basic human needs. The dryland villages are less progressive in terms of globalization due to their remote location in unproductive arid and semi-arid landscapes where tribal communities predominantly inhabit the wilderness. Senior government officials seldom venture out to explore these remote areas due to logistical hassles.

India's drylands in general receive less rainfall when compares to other fertile river basins so droughts tend to plague the region during weak monsoon periods. Therefore, saving every drop of rainwater has been a tradition for survival in the harsh wonderland where people still remain tough, physically and mentally, somewhat resembling the rugged backdrop.

Humanity cannot sustain without freshwater since it is an essential entity to quench thirst. The question however is, can water reduce rural poverty?

India's unique rural development pundit, Harnath Jagawat, passionately supported by his compassionate wife and colleague, Sharmishtha Jagawat, a prominent social worker, started to use the magic of water nearly four decades ago to mitigate poverty in drylands. Reinforced by an army of dedicated employees that include irrigation engineers specialized in building lift irrigation systems and check dams, technical experts specialized in agriculture, horticulture, floriculture and vegetable farming, and many village-level social workers, the Jagawats initiated poverty eradication work four decades ago. They still continue to build series of cost-effective minor irrigation infrastructures across the drylands of western India to provide desperately needed irrigation water to farmers since big dams connected to canals cannot reach the undulating farms inhabited by the widely scattered tribal communities.

Till March 2016, a total of 406 lift irrigation structures and 385 check dams were constructed in tribal drylands, which in turn converted over 110,000 acres of wasteland to productive agricultural land through community irrigation cooperatives. This rejuvenated green revolution pioneered by the Jagawats intertwine various farm-based income generating activities that ultimately benefitting over two million farmers in tribal villages of western India. These families lived in poverty prior to the intervention. Now, each family saves an average of over ₹10,000 yearly and they have finally come out of poverty and hunger.

Finding freshwater in India's drylands is a luxury. But, the Jagawats have succeeded in harvesting water using check dams and then lifting the water from low-lying streams, rivers, ponds, and check dams to transport it the upland farms. They have followed Mahatma Gandhi's vision of village-level development to sustain life by using the magic of raindrop. Traditionally, tribal farmers migrate to cities in search of work since they could afford only rainfed farming. When monsoon fails, they suffer the worst in terms of poverty and hunger. After the construction of lift irrigation systems and series of check dams, irrigation cooperatives were established in villages, crop productivity increased, and at the end the village-level poverty was replaced by food security. The farmers now have no need to migrate to cities since access to irrigation water is no longer an issue. Irrigation cooperatives manage the infrastructures and farmers pay a modest sum for water to irrigate their dryland farms.

The grassroot-level sustainable development established by the Jagawats involve lift irrigation, check dam, renewable energy, horticulture, floriculture, dairy farming and other income generating farmland activities directly and indirectly contributing to food security and poverty reduction, both local and national levels. As showed in this book, transforming India's vast unproductive drylands to achieve a contemporary green revolution to increase food security and to decrease poverty is certainly possible with the support of the Sadguru Model of Rural Development pioneered by Harnath and Sharmishtha Jagawat.

Govindasamy Agoramoorthy, Ph.D.
Distinguished Research Professor
Tajen University, Taiwan
Tata Visiting Chair
NM Sadguru Water and Development Foundation
Dahod, Gujarat, India
Research Advisor
SVYAS Yoga University, Bangalore, India

Contents

Chapter 1
Beginning of a Legacy

"Nature is infinitely creative. It is always producing the possibility of new beginnings"— M. Williamson

Introduction

Among all the countries that I travelled and lived over the last half century, I find India the most fascinating of all. Though I was born in the south Indian state of Tamil Nadu, I was attracted to Rajasthan's vibrant culture and heritage from the college days. As a matter of fact, my association with Rajasthan started long ago in 1982 when I was invited by the renowned naturalist, professor SM Mohnot (he goes by SM) to do doctoral studies at Jodhpur University. He also offered a senior research fellowship. When I received SM's invitation to conduct field research on the intriguing Hanuman langurs at Jodhpur, I was working as a research assistant to India's ornithology legend, professor Salim Ali. Though my job was to monitor migratory birds at Point Calimere wildlife sanctuary, I spent more time to watch the common monkeys, the bonnet macaque. Monkeys fascinated me from childhood for reasons that I still cannot comprehend.

When I was about to leave my village, my father asked, 'why do you go all the way to Jodhpur located near Pakistan to study monkeys when they come to your backyard?' He pointed out to a group of bonnet macaques that used to visit my farm daily. But I hardly had an answer to his odd query and faded away. I took a long train journey from Chennai to Jodhpur. While looking out the window, I was mesmerized at the enormity of India's diverse landscape. After reaching Jodhpur, I enjoyed observing wildlife and got immersed in local culture. Though the desert culture and wildlife intrigued me a lot, I was charmingly puzzled by the magic of water in the desert.

Magic of Water

When I was little, my family used to get water from our house well. A hand pump likewise sat on the front yard that had water year-round. A tributary of the holy river Cauvery ran behind my farm so the village remained green constant with ample of water. Whenever I drank water from the clay pot, it was cool and delightful. Hence, I thought that people everywhere had wells and pumps that held all the water they would ever need. While I lived on the outskirts of Jodhpur to observe wildlife, I saw women walking for miles to fetch water. I also ended up walking with a bucket for a mile daily to collect water. At Jodhpur, I learnt the art of taking bath with a single bucket of water. My stay at Jodhpur reminded me what the writer Vera Nazarian wrote long ago, "In the desert, the only God is a well". I didn't know the value of water when I was in Tamil Nadu and I didn't feel guilty of wasting the precious drinking water. I started to revere the water only after landing in the arid landscape of Jodhpur.

Likewise, the American naturalist and philosopher Loren Eiseley wrote, "If there is magic on this planet, it is contained in water". But, I didn't comprehend the magic until I reached Sadguru Foundation's office in Chosala village (Dahod District, Gujarat) during the summer of 2006. For over four decades, the foundation has pioneered rural development work to relieve poverty among the impoverished tribal communities. When I first met the distinguished rural development leaders, Shri Harnath Jagawat and Smt Sharmishtha Jagawat (Directors of Sadguru Foundation), I was instantly fascinated by their dedication centered on irrigation water. Since then, I have visited several field sites where they are actively promoting community-based development work. I also spent long hours talking to the Jagawats to better understand the complex issues revolving India's rural development work.

The Jagawats have established an excellent system to harvest rain water using check dams in small, medium and large rivers. They have dutifully followed what the national father, Mahatma Gandhi highlighted on the philosophy of village-based business. They have rigorously tested Gandhi's economic principle using the power of water as a basic tool to reward life.

Traditionally, tribal farmers used to grow one seasonal crop based on the often meager rainfall. Droughts plague the drylands every few years and the harsh reality forces people to migrate to nearby cities for work. Ever since the Sadguru Foundation established check dams and lift irrigation systems to irrigate farm, migration to urban areas in search of labor jobs stopped, crop productivity increased and poverty eradicated.

Mahatma Gandhi understood the concept of rural development long before the catch phrase "sustainable development" came into existence in literature. Before his death, Gandhi prescribed plans for "total development" of society through non-violence in harmony with nature. His ideas of economics are unrealistic, as few could practice his way of life, especially his preference for simple types of production activities. But, there has been renewed interest in his economic ideals since the materialistic way of development has led to serious ecological crisis affecting community harmony. Gandhi's prescription of "total development" involves the fulfillment of material needs not the only aspect of development, but also the need for nature protection without undermining the ethical values in society.

Gandhi's perspective spins around the values of society and nature, which blends economics, ethics and morality rather equally. In fact, it's far better than the characteristics portrayed by the contemporary catch phrase "sustainable development". Moreover, the Gandhian perspectives are consistent with the ancient concept of *Arthasastrta* that encompasses a wide range of societal issues. By firmly believing Gandhi's ideology, the Jagawats initiated rural development by showing tribal farmers how to harvest rainwater using traditional infrastructures. Their efforts have so far benefited over two million tribal and non-tribal farmers, who live in the impoverished drylands of India.

Humble Beginning

Harnath Jagawat grew up in a remote village called Ekkalgarh located on the bank of river Chambal in India's Madhya Pradesh state. I visited his village few years ago and it still remains remote where the landscape influenced by the reddish and rocky backdrop mimics the planet Mars. Chambal, by the way is India's only river that has the status of wildlife sanctuary because of three critically endangered species namely Gharial (fish-eating crocodile), Red-crowned roof turtle and Ganges river dolphin that depend on the wild river for their survival. Although the village had immense natural resources, poverty was rampant during childhood of Jagawat. So, he wanted to find solutions to enhance socio-economic conditions of Indian village. This shows how people tend to develop childhood desires catalyzed by their natural surroundings. The deep-rooted craving for social service targeting the underprivileged motivated Harnath to pursue his master degree in social work. While studying at the MS University in Vadodara city, Harnath met Sharmishtha, who had similar views of serving the tribal society. Since then, the social work couple dedicated their lives to serve people, especially the neglected tribals inhabiting India's drylands. Right after graduation; Harnath became lecturer in a training institute. Though he liked the training duties, his inquiring mind serving the impoverished people propelled him to look for an opportunity for action elsewhere that naturally came to him in 1974.

Water-centered Charitable Trust

India is home to the Indus Valley Civilization with a long history of global trade, great empires and unique cultural wealth. So, entrepreneurship is nothing new for India though the business has come a long way. India has over 100 billionaires now and many Indian-owned companies are listed in Fortune 500. Nevertheless, I find the entrepreneurial history of Shri Arvind Mafatlal fascinating and unique since he belongs to a tiny group of unusually compassionate tycoons who saw the spirit of God while serving the poor.

Figure 1. Shri Arvind Mafatlal and his wife, Smt. Sushila Mafatlall with their guru, Shri Ranchhoddasji Maharaj in 1968 (Photo courtesy Shri Hrishikesh Mafatlal).

The Mafatlal Group has been a leader in India's textile business for over a century. Hence, I am familiar with the brand names like Mafatlal, Tata and Birla from childhood. Though I never had an opportunity to observe him in action, historical records show that Shri Arvind Mafatlal as an amazing entrepreneur, who was more empathetic to share his wealth to help the needy suffering from extreme hunger and poverty.

Arvind Mafatlal had great reverence towards his guru, Shri Ranchhoddasji Maharaj. Based on his inspiration only, Mafatlal founded the Sadguru Seva Sangh Trust in 1968 as a charitable organization to assist India's poverty-stricken communities (Figure 1). He dedicated his life serving humanity by organizing numerous social service camps throughout India while progressing well with his family textile business.

Serving people in India's remote villages and simultaneously running an international business venture was not at all an easy task in those days. But, Arvind Mafatlal somehow managed to achieve both the difficult missions gracefully and efficiently, like burning a candle on both sides. He diligently worked with all social work camp operators and even served food for workers and villagers on numerous occasions (Figure 2).

In fact, Mafatlal's compassion was not limited to humans. When the Kachchh region in Gujarat state suffered serious droughts during the early 1970s, large number of cattle started to die off due to starvation. Mafatlal was there to rescue them from hunger and death. He built cattle shelters and provided them necessary food, water and care. Whenever natural disasters struck India through droughts, cyclone and earthquakes, Mafatlal was there personally to serve humanity and to relieve human suffering. He was so low profile and most people could not even notice his presence as he worked tirelessly serving people. It evidently shows that Mafatlal saw himself in all lifeforms, which is an extraordinary human quality. It reminds me of a statement I studied in Katha Upanishad that states, "Those wise ones who see the consciousness within them is the same consciousness within all beings, attain peace".

Figure 2. Shri Arvind Mafatlal serves food for people affected by natural disaster (Photo courtesy Shri Hrishikesh Mafatlal).

In 1974, Mafatlal organized an eye camp at Dahod (Gujarat state) to help the tribal people. At that time, the Jagawats were given an opportunity to bring patients from far away villages in Gujarat and neighboring Rajasthan and Madhya Pradesh states. Nearly 2,000 patients were transported back and forth from far away villages. The Jagawats worked hard to make the event a great success. Mafatlal was impressed with their hard work and dedication. All patients who attended the eye camp were checked and treated by expert doctors promptly. The patients were also given free meals.

What intrigued Mafatlal was that people were more eager to get free meals than checking their eyes. So, he wondered why? He instantly requested Harnath to explore the mystery surrounding his observation. Then, Harnath spent the next few days talking to tribal farmers and carefully studying the ground reality by visiting villages. He found that the farmers had lands to cultivate; they could afford only rainfed farming depending on monsoon rains. But, droughts plagued the area that forced farmers to leave their homes in search of food and work elsewhere. That was the reason farmers who came to eye camp were more eager to get free meals than checking their eyes. Hunger was widespread among farming communities in those days. So, Harnath came up with a unique idea to provide irrigation water to solve the issue of hunger and poverty.

Although drylands receive fewer rains, they do have several small, medium and large rivers crisscrossing the rugged terrain intertwined with farmlands. The rivers store water for several months after the end of the monsoon season. But, rivers are often located in deep valleys while farms go upwards so no way to use gravity for irrigation. In other words, farmers who live up the hill can see enormous water stored deep down in river valley, but cannot afford the technology to lift water.

Harnath therefore decided to set up the lift irrigation systems using electric motors to pump water from rivers to irrigate the upland farms. Sharmishtha took up the role to communicate with tribal women to mobilize them as groups across villages. With the blessings and support of Shri Ranchhoddasji Maharaj and Shri Arvind Mafatlal, the Jagawats were able to initiate the long march towards rural development with the creation of NM Sadguru Water and Development Foundation (popularly known as Sadguru Foundation) in 1974 to eradicate hunger and poverty in tribal villages of western India.

Figure 3. Shri Hrishikesh Mafatlal with his father, Shri Arvind Mafatlal (Photo courtesy Shri Hrishikesh Mafatlal).

The current Chairman of the Mafatlal Industries, Shri Hrishikesh Mafatlal continues the family legacy of helping the underprivileged, and he strongly supports Sadguru Foundation's rural development activities in villages (Figure 3). He respectfully attends Sadguru Foundation's Board Meetings and carefully monitors the progress of social work centered in tribal areas of western India.

Figure 4. A recent photograph of Shri Harnath Jagawat and Smt Sharmishtha Jagawat taken at Udaipur during staff retreat in 2016 (Photo by S. Chaudhary).

I observed Hrishikesh during a recent board meeting of Sadguru Foundation at Ahmedabad and was puzzled by his sharp rational query. He asked the staff whether Sadguru Foundation has any data to show how rural development work minimized farmer's debt and maximized income and savings in villages. The logical inquiry instantly reminded me of his father's vision to start Sadguru Seva Sangh Trust to eradicate poverty in tribal villages. Hrishikesh genuinely continues to focus on his father's social work ideology mandated for Sadguru Foundation. More details on the history of Arvind Mafatlal's compassionate legacy of serving the Indian society can be seen in the book titled "Shri Arvind Mafatlal: A life lived with grace" elegantly compiled by Shri Hrishikesh Mafatlal.

Work ethics catalyzed by the compassionate legend, Arvind Mafatlal and his spiritual sage, Shri Ranchhoddasji Maharaj, the Jagawats devoted their adult lives for karma yoga serving India's village communities. Karma yoga literally means selfless action as a way to perfection without ego. The Jagawats continue their efforts to assist the impoverished tribal communities by providing irrigation water that eventually eradicates hunger and poverty.

When people ask the Jagawats how many children they have; they humbly reply that they consider all the tribal people of India as their children. They have enormous enthusiasm for social work targeting the underprivileged to enhance livelihood opportunities. Although the Jagawats are now in their 'early 80s' (Figure 4), they are still very active, and they always keep their minds tuned to the latest ground reality on the rural development agenda.

When I asked Sharmishtha, when the formal training of farmers started? She replied that when they started the lift irrigation schemes, it was a new technology those days, so they ended up in teaching farmers via field exposure trainings to show them on how to start or stop machines, handling procedures, and reparation issues, etc. It was all new for the tribal farmers since they have never used machineries in traditional agriculture. The formal training programs started in 1992. By the way, Sadguru has been doing trainings for farmers since the beginning by explaining about their mission. Initial trainings were held in open fields under trees and then under tents.

Figure 5. Annual gathering of lift irrigation farmers organized by the Lift Irrigation Federation and Sadguru Foundation at Andeshwar in Banswara district of Rajasthan state (Photo courtesy Sadguru Foundation).

When the Jagawats started formal and organized training courses, other NGOs, government agencies and other farmer groups started to meet them to learn about their community organizational strength. Therefore the training institute was developed and later established with fantastic infra-structure. I have personally seen the efficiency of Sharmishtha in mobilizing villagers and she can get thousands of women for a meeting in just one phone call (Figure 5). When Sadguru Foundation started their work, Harnath just had an old scooter and there was no electricity in Chosala village.

Sadguru's Community-based Development

Sadguru Foundation's corporate support initially came from Mafatlal and then from the Tata trusts (Sir Ratan Tata Trust, Sir Dorabji Tata Trust, etc.) to implement various rural development schemes in tribal areas to purge poverty. Local government agencies in Rajasthan, Gujarat and Madhya Pradesh states continue to request Sadguru to build more check dams and lift irrigation systems in remote areas; they are now offering 100% grants with about 10% administrative overhead, which shows Sadguru Foundation's transparency and clean governance. It has indisputably developed a fascinating model to amicably work with both government and corporate sectors— other government agencies and NGOs can follow suit this unique strategy.

During my visit to villages, I have witnessed how water could clean up the fraudulent lifestyle of people in some villagers. One such example worth to mention happened in a small village called 'Karmakhedi' (Jhalawad District, Rajasthan). The name of the village "karma" originated long back when the tribal community indulged in unlawful endeavors such as banditry and producing illicit liquor. When check dams were built near the village, people started to irrigate their fields through lift irrigation cooperatives that abruptly stopped their illegal activities. The women group in the village now wants buffalos and they unanimously said that Sadguru Foundation taught them to earn better and to upgrade their socio-economic status. When I asked what their future dream would be; the answer was to educate children and to start other cooperatives to earn more to improve livelihoods. A 45-years old mother, Rasalbai whispered, "Sadguru taught me truthfulness and honesty', and she added, 'everyone should do their part to help others".

Gandhi's dream of grass roots political participation is likewise happening in villages of Rajasthan. Each and every community irrigation cooperative established by Sadguru Foundation now includes several politicians that include village council heads and members, and even taluk level representatives with a potential to reach legislative assembly and parliament in near future. Those who have been elected as political representatives have in fact worked with Sadguru for 4 to 5 years and it shows Sadguru Foundation's capacity building contributions to community members transforming them as leaders in local politics.

Political leaders in Gujarat, Rajasthan and Madhya Pradesh states realize the impact of Sadguru Foundation's presence in tribal areas and are now eager to reach out to remote tribal villages, which is good news. However, Sadguru Foundation maintains the status quo of not interfering in political affairs. But, the grassroots-level irrigation cooperatives have already been enlightened with political change. So, the bottoms up approach might be the best way to revive the Gandhian perspectives of political and economic change in India's villages.

Sadguru Foundation's Project Area

NM Sadguru Water and Development Foundation, also known as Sadguru foundation, is a nonprofit organization based in Chosala village of Dahod District, Gujarat State (Figure 6). It has been working in over one thousand villages spreading across Gujarat, Rajasthan and Madhya Pradesh states of western India. The project area is classified as drought-prone semi-arid region in India. The region is predominately inhabited mostly by the impoverished tribal communities that are the weakest section of the Indian society.

The project area now extends in sixteen districts of Rajasthan, Gujarat and Madhya Pradesh. Besides implementing livelihood programs centered on water and natural resource management, Sadguru Foundation has been largely imparting training, capacity building and technical inputs to large numbers of government and non-government organizations at its state of the art training facility at Chosala village.

Most non-profit agencies operate out of cities and often use the weaknesses in the government and corporate system to condemn mostly on the inadequacies and shortcomings. But Sadguru operates from the tribal village of Chosala. During working days, all staff members including the Jagawats gather at about 8:15 am daily in front of Sadguru Ranchhoddasji Maharaj's statue and start their work by chanting the following prayer:

"O, the most compassionate supreme deity,
We want to pursue your mission that you started long ago,
Bless us with strength and courage to complete the mission,

Make us work with love, harmony, worship and friendship,
Kindly watch our honesty and sincerity towards the work,
Forgive our mistakes and help us to rectify them promptly,
Bless us to perform our duties with perfection".

PROGRAMME AREA - STATES OF SADGURU FOUNDATION

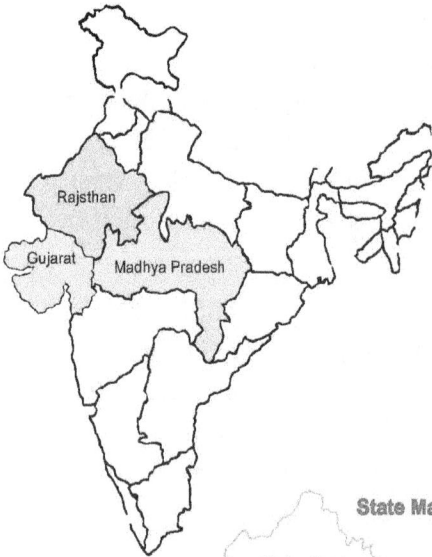

Rajsthan
Gujarat   Madhya Pradesh

State Map

Rajasthan

Gujarat   Madhya Pradesh

District Map

Kota
Jhalawar
Mandsaur
Pratapgarh
Dungarpur   Banswara
Sabarkantha
Ratlam
DAHOD
Panchmahal's   Jhabua
Vadodara   Alirajpur

MISSION

SADGURU endeavors to develop and expand environmentally, technically and socially sound natural resource interventions leading to empowerment of rural community including women to ensure equitable and sustainable development and poverty reduction.

VISION

Empowerment of tribal and rural communities with natural resources restored, developed and expanded in the selected project areas.

Figure 6. The project areas where Sadguru Foundation implements rural development work in India' drylands.

Afterwards, a short briefing of senior managers is held in the meeting room where department heads share their past day's work and explain their plan of work for the day. Department coordination meetings happen monthly where all staff members are briefed on the achievements and operational aspects of all projects. All Sadguru staff members, with whom I interacted, shared their experiences passionately. I am puzzled by the fact that Sadguru's engineers can build even large check dams in a matter of months. They work with basic equipments and monitor construction work round the clock. This unusual dedication and simplistic way of functioning can come only with genuine enthusiasm and deep compassion to help the poor. Check dams built by Sadguru three decades ago still stand intact, like the Grand Anicut, demonstrating the experience, integrity and professionalism of staff. I was annoyed to see many other check dams built by government contractors in Rajasthan, Madhya Pradesh and Gujarat states disintegrated and remnants of such check dams can still be seen as tombstone in the barren landscape. The Sadguru model of using the magical water to sustain life through check dams is an option that India can no longer ignore.

People need to develop social constraints on natural resources use if their societies to survive. And deciding what, where and against whom constraints are necessary, and how they are to be instituted presents serious dilemmas. As the human population has gone above Earth's ecological carrying capacity, issues of both inter-group and inter-generational equity, and the intimately connected ethics of the treatment of human life support ecosystems are now moving to the forefront, and will be among the greatest moral concerns of the 21st century.

It's not too late for humanity to avert vast ecological disasters by making a transition to sustainable living, but the test will not be simple. Wasteful consumption in rich countries must be reduced to allow needed growth for poor countries. For instance, physicist John Holdren's scenarios, in which the rich become much more efficient and the poor consume more, offer a possible path towards more equitable and efficient form of energy use that could close the gap between rich and poor and reduce environmental damage compared to the continuing current trends that broadens the global gap between rich and the poor (Ehrlich, *et al.*1978). The cooperation that will be needed to solve global environmental problems is unlikely to be achieved in a world divided into rich and poor and driven by ethnic and economic antagonism. It is therefore essential to choose appropriate sustainable development projects at the grassroots in villages to root out poverty and enhance natural resources management in our egion and beyond.

Figure 7. Harnath Jagawat dressed in the traditional outfit to attend meeting of community leaders (Photo courtesy Sadguru Foundation).

Sadguru Foundation's programs in villages through its pragmatic leadership and vision have given employment to over two million tribal people with a gross income of thousands of crores of Indian rupees annually through agricultural productivity. Harnath Jagawat (Figure 7) says that even large government agencies and conglomerates won't be able to pay roll ten lakhs employees at one time.

The economic impacts of Sadguru Foundation go far beyond several folds since those who benefited directly and indirectly started to replicate the derived ideas to help others. Above all, the often marginalized tribals, the guardians of natural resources, regained the self-confidence and dignity at last.

The Sadguru Model of rural Development pioneered by the Jagawats continues to relive poverty and enhance food security across thousands of villages in the tribal dryland regions of western India for over four decades. So, this model has great potential to be replicated across other regions of India and elsewhere. Jagawat has written numerous articles including books on his rural development journey and they can be accessed at www. nmsadguru.org. Some selected papers and books are listed below in the references section at the end.

Chapter 2

India's Largest Check Dam in Rajasthan

"Wild rivers are earth's renegades, defying gravity, dancing to their own tunes, resisting the authority of humans, always chipping away, and eventually always winning" — R. Bangs

Introduction

Sadguru Foundation has focused mainly on providing irrigation water to tribal farmers. Hence, it started to build lift irrigation structures to pump water from low-lying rivers and reservoirs to upland farms. Later, the focus switched to build series of check dams across all minor and major rivers in drylands (Jagawat, 2005). When I first saw the mega check dam on river Mahi in Rajasthan, it reminded me of the Grand Anicut located in my home state, also known as the 'Kallanai' in Tamil language. It was built during the 2nd century AD by the Chola King, which is still supplying water to the Cauvery delta in Tamil Nadu.

The holy river Cauvery, also known as *'Daksina Ganga'* in Sanskrit (Ganges of the South), has been treasured by the people for centuries. It is called the lifeline since it irrigates vast areas in the delta and four million farmers are entirely dependent on the river for irrigation water. Sharing the Cauvery river water has been an issue of contention between Karnataka, Tamil Nadu, Kerala and Pondicherry states for decades. The Cauvery Water Disputes Tribunal was specifically set up in June 1990 to deal this issue. The tribunal gave its legal verdict on sharing of water among the four riparian states. The tribunal has also recommended that the central government constitute a Cauvery Management Board to implement the directions in its final award. It is crucial for this body to initiate a new task to harvest the monsoon flow in the delta region.

After seeing the largest check dam in Rajasthan, I was convinced that Tamil Nadu can solve the river water sharing crisis with Karnataka if the state could build series of check dams in Cauvery and its tributaries throughout the delta (Agoramoorthy and Hsu, 2016). Sadly, there is not much political and bureaucratic interest for micro-irrigation projects since they believe in large irrigation projects involving huge amount of cash. In fact, there is no place to build any large irrigation infrastructures in Tamil Nadu. Besides, people are traditionally used to seeing water and even floods from childhood, so, they simply cannot comprehend the concept of rainwater harvesting.

Baneshwar Dham Check Dam

When I saw the large Baneshwar Dham check dam in Banswara District of Rajasthan, I was overwhelmed. That was the first time I encountered an eco-friendly big-sized check dam that neither displaced people nor destroyed nature. When it opened in 2007, it had a price tag of one million US dollars. The 367 meters long and 7.3 meters tall dam expand 7,000 acres of irrigated area, ultimately benefitting over 18,000 local farmers (Figures 1 and 2).

The Honorable Chief Minister of Rajasthan, Smt. Vasundhara Raje inaugurated the check dam. The dam has water storage capacity of 350 mcft. The Chief Minister appreciated Sadguru Foundation's enormous contributions to rural development through construction of large check dams in the water-scarce Rajasthan state. The government had built a bigger Mahi-Banas Sagar dam on the upstream. River Mahi originates in the Mahi Kanta hills in western Madhya Pradesh, and enters near Chandangarh in Rajasthan's Banswara District.

As a matter of fact, the Sadguru Foundation has built several large check dams across the arid and semi-arid regions of Rajasthan for many years. The mission was possible because of better cooperation from the government of Rajasthan. Several high-level officials, both from political and bureaucratic circles, continue to show great interest in check dams since they fully understand the significance of eco-friendly structures benefiting society and enhancing preservation of ecologically sensitive drylands in Rajasthan

Figure 1. India's largest check dam built by Sadguru Foundation on river Mahi at Baneswar Dham in Banswara district in Rajasthan (Photo by G. Agoramoorthy).

Figure 2. Baneswar Dham check dam's back water goes long distance can be seen clearly in this photograph (Photo by G. Agoramoorthy)

The following photographs show one such large check dam located in Jhalawar district of Rajasthan, before and after construction, and how the landscape has been hydrated by the eco-friendly small dam (Figure 3).

Figure 3. A view of the dried out river Chambal near Sindhla village in Rajasthan state during January 2002 before check dam construction (above) - a view of the same spot during November 2003 after building check dam where backwater go for miles (below) showing the enormous potential of small dams (Photo courtesy Sadguru Foundation)

Sadguru's Unique Rural Development Scenario

It is tough for any non-profit agency to implement check dams, lift irrigation systems and other projects since huge sum of money is needed to implement projects. Though the Jagawats faced hardship for funds initially, they were able to develop a unique model of transparency in working with government, non-government and corporate sectors to implement various agricultural-based development schemes in villages. When the Jagawats approached government agencies in Gujarat, Rajasthan and Madhya Pradesh with workable plans of lift irrigation systems and check dams, sponsors followed them to support. Traditionally, government contractors used to carry out minor-irrigation projects that often ended in failure. Similarly, corporate support has been generous for Sadguru, initially from the Mafatlals and afterwards from the Tatas, mainly through Sir Ratan Tata Trust and Sir Dorabji Tata Trust. Besides, other agencies such as the Ford Foundation, Aga Khan Foundation, Norwegian Agency for Development, Axis Bank, National Bank for Agriculture and Rural Development, Mahindra and Mahindra Limited, Coca-Cola India, and many others came forward to support rural development work in drylands. Thus, a small non-profit agency based at a remote tribal village located along the borders of Gujarat, Rajasthan and Madhya Pradesh indisputably developed a fascinating model of corporate social responsibility with an effective strategy to amicably work with both government and private sectors. Other national, regional and international agencies should follow suit this classical textbook model of development with transparency.

India's ruling elites have a long history of considering government funds as their own leading to mismanagement. The success of Sadguru's watershed management work is simply based on transparency, accountability and responsible utilization of public funds. Besides, Sadguru Foundation used its unique technical competency to implement irrigation infrastructure projects by removing ambiguities and ensuring uniform interpretation of government's implementation guidelines while avoiding contradistinctions in policies of different agencies. While harvesting water using check dams and pumping water through lift irrigation systems, the foundation initiated several land-based income generating activities that includes agriculture, horticulture, floriculture, agro-forestry, farm forestry, diary farming and so on in the drylands covering numerous villages. The Jagawats have given importance to human resource development and institutional capacity building procedures to empower both men and women farmers to initiate and implement various sustainable agricultural development projects on their own. Hence, the Jagawats have adopted various working strategies that include awareness creation, training of villagers, capacity building to empower women, and building institutions.

Thus far, Sadguru has astonishingly created thousands of village-based peoples' institutions such as lift irrigation cooperatives, watershed associations, self-help groups, youth and farmers' groups, women horticulture groups, and joint forest management cooperatives. It has an excellent training campus with dormitory facilities where hundreds of local farmers and overseas visitors get hands-on training. In fact, it has already provided various types of trainings to government agencies in the fields of irrigation, water management, and agriculture. For example, leading civil administrators from the Indian Administrative Service (IAS) took part yearly in training at the foundation to understand poverty-reduction schemes. Besides, officials of all categories including the National Bank of Agriculture and Rural Development or NABARD, a major sponsor of India's rural development work, continue to participate in training at the foundation's headquarters each year. Sadguru's training center with guest houses, dormitories, conference halls, and meeting rooms provide better opportunities for hands-on learning about rural develpment.

Figure 4. Rajasthan Chief Minister visiting Sadguru Foundation in 2016 (Photo courtesy Sadguru Foundation, India).

The Jagawats believe that both government and non-government agencies must co-exist, cooperate and support to achieve the goal of sustainable development through social work. They have created trust among inter-state government agencies. Consequently their projects ranging from check dams to lift irrigation, and agriculture to solar lighting in tribal villages across western part of India rapidly multiplied in recent decades. Astonished by Sadguru's social work, the Chief Ministers of Rajasthan (Figure 4) and Gujarat have visited villages where Sadguru Foundation implemented field pojects.

Figure 5. A large check on river Kshipra in in Parasali village (Jhalawar district, Rajasthan) and it stores 75 mcft of water, supporting 9 lift irrigation schemes and large number of potable pumps irrigating 1500 acres. Four large check dams on this river are located within 25 km range revived the river with water most of the year (Photo courtesy Sadguru Foundation).

The Baneswar Dham check is an example of simple innovative technology to harvest water and boost agriculture. The Jagawats have understood India's ingenious past in the eco-friendly dam building saga. With the help of dedicated engineers who are specialized in building dams, he has boosted agricultural productivity in tribal areas by means of cost-effective technology (Figure 5). About 70% of India's tribal population is in concentrated in eight states such as Bihar, Jharkhand, Odisha, Madhya Pradesh, Chhattisgarh, Gujarat, Rajasthan and Maharashtra. These areas are the most backward in terms of economic development, therefore poverty is rampant. The Sadguru model of using water to sustain life through check dams and lift irrigation systems in these areas can be easily done with the government support.

Historic Check Dams

The classification of dam has been centered on its physical appearance, mainly height since it plays a major role in furthering the backwater volume.

According to World Commission on Dams (2000a, b), when the height of a dam goes larger than 15 m, it becomes a large dams; when it tops 150 m, it is branded as major dam. China's Jinping-I is the tallest (305 m) major dam in the world. Despite its height, the Jinping-I is not all that magnificent since the Hoover dam (221.4 m tall) is still admired as the exceptional engineering spectacle of the 20th century.

When the height of a dam shrinks below 15 m, it becomes a small or check dam; they are built across streams and rivers to harvest the rain drops. In other words, check dams are small barriers using stones, cement, and concrete built across the direction of water flow on rivers to harvest rainwater in reote areas.

Figure 6. China's oldest irrigation system (above) at Dujiangyan in Sichuan province (Photo by G. Agoramoorthy); India's oldest Grand anicut in river Cauvery (below) at Tiruchirapalli, Tamilnadu state (Photo courtesy en.ikipedia.org).

As a matter of fact, check dams have been around for millennia across the world in ancient cultures. The evidence can be visible in the historic ruins of the Mayan civilization in North and Central America, starting from Mexico to Guatemala and El Salvador to Costa Rica (Luzzadder-Beach *et al* 2012).

Archeological records show proofs of check dams from the Indus valley civilization that once flourished from India to Egypt (Albinia 2010). Ancient cultures in the Arabian Peninsula, Sinai desert, Yemen, and Jordan also had check dams to harvest rainwater (Mccorriston and Oches 2001; Beach at al. 2002). The eco-friendly edge of check dams might have perhaps extended their longevity. For example, the Grand Anicut (height 4.5 m) built during 2nd century AD in India's sacred river Cauvery still functions flawlessly. It is the oldest check dam built by the Chola dynasty to promote rice cultivation in the fertile delta. Similarly, China's earliest irrigation structure at Dujiangyan included check dam and diversion canal and it was built in 256 BC near Chengdu in Sichuan province (Figure 6). It reduced floods in the Min river delta and expanded irrigation for centuries (Cao, *et al.* 2010).

During the British occupation of India, two military engineers pioneered modern large-scale irrigation: Proby Cautley, who built the Ganges Canal, and Arthur Cotton, who rebuilt the Grand Anicut on the Cauvery—both systems of redistributing water over hundreds of miles of canals (Figure 6). In 1799, when the British East India Company took control of the Cauvery delta, it was unable to check the rising river bed due to silt backed up against the dam. Company officials struggled for a quarter century and finally, using indigenous technology, Arthur Cotton was able to solve the problem by renovating the Grand Anicut. He later wrote: "It was from them (Indians) we learnt how to secure a foundation in loose sand of unmeasured depth. The Madras river irrigations executed by our engineers have been from the first the greatest financial success of any engineering works in the world, solely because we learnt from them (Cotton, 1874).

Farmers Struggle for Irrigation Water

In 2008, I visited the Alawa village that has a population of 1000 people and it's located in Jhalawar District of Rajasthan to understand the issue of irrigation water scarcity. Armed with diesel engines, farmers were seen aggressively pumping water from two check dams located in the Ahu River, a tributary of Kali Sindh, which originates in the northern slopes of Vindhya in Madhya Pradesh. Despite attempts over two decades by the farmers to persuade the panchayat to build check dams in the Ahu River, their efforts have failed. However, in 2006, the farmers approached Sadguru Foundation and explained about their desperate need for irrigation water. Subsequently, with the support of the Government of Rajasthan and matching funds from Sir Ratan Tata Trust and Sir Dorabji Tata Trust, the Sadguru Foundation was able to build two check dams namely Alawa-I (length 179.55 m, height 2.97 m) and Alawa-II (length 249.35 m, height 2.47 m) with a cost of rupees 75 and rupees 82 lakhs, respectively. As a result, the irrigated area around the Alawa village increased by 500 acres, due to the availability of 25 mcft of water stored by two dams ultimately benefiting 107 households or 650 people.

By seeing the check dams, farmers from neighboring villages started to compete for the same water source. I was astonished to witness farmers placing hoses for 3-6 km to transport water from the river (Figure 7). The intense competition for irrigation water resulted in draining even the last drop of water from the check dams and some farmers even dug wells in the already dried river Ahu in pursuit of grounwater (Figure 7).

Figure 7. Irrigation water competition in river Ahu, Rajasthan state (above) where farmers use diesel engines connected with hoses to transport water for long distances to irrigate farms. A view of the completely dried up check dam due to intense competition for irrigation water in Alawa village with a newly dug well in the midst of the river (Photo by G. Agoramoorthy).

It clearly shows how desperate the farmers can become to access irrigation water, especially in the semi-arid regions of Jhalawar District, which is one among the most impoverished and least developed districts in India. Intrigued by the irrigation water saga, I participated in a community meeting in the village to discuss options to solve the irrigation water standoff (Figure 8). I was awestruck when the farmers came up with the following five major strategies to mitigate the water criis that include:

Figure 8. Community meeting to sort out the irrigation water crisis in Alawa village in Rajasthan state (Photo by G. Agoramoorthy).

1) source for low water consuming crops, 2) set up regulations to restrict the use of water by farmers from adjoining villages, 3) set up a committee to minimize irrigation water wastage, 4) reduce the usage of private diesel engines and replace with electrified lift irrigation system managed by village cooperative, and 5) discuss with Sadguru Foundation's directors and engineers to increase the height of check dams to boost water storage.

It showed how the rural farmers could handle water crisis more rationally and scientifically using their invaluable practical knowledge of natural resources management. In fact, the farmers' strategic plan to deal water crisis in Alawa village coincide with the recent comprehensive scientific assessment on water management in agriculture compiled by the International Water Management Institute that calls for radical changes to the way the world produces the food and manages the environment (Molden, 2007.).

Later, I discussed this case with Sunita Chaudhary, Sadguru Foundation's lead civil engineer involved in the construction of numerous check dams and she said that it would be possible to slightly increase the height of check dams without drowning nearby farms and damaging the river bank. She reiterated that numerous villages across western India face similar water shortages for irrigation and building more check dams with the support of government, NGOs and private corporations as the best way to ease future irrigation water conflicts in rural areas.

India will require over 300 million tons of grain to feed its people in near future, but productivity is declining in recent years. Understanding this grim situation, India's Prime Minister has increased the spending to 36,000 crore in the 2016 budget allocation to diversity rural agriculture, which is a pleasant news for farmers. Likewise, the Chief Minister of Rajasthan State is also keen to diversify agriculture and that's why she is eager to build series of numerous large check dams in rivers.

Chapter 3

Environmental Benefits of Check Dams

"A man of wisdom delights in water"—Confucius

Introduction

Drylands spread across nearly 40% of our planet's surface and they dominate in Africa, followed by Asia. Scientists have predicted that desertification will intensify from Kansas to California in the United States, so even any well-developed nation won't be spared by climate change aftermaths in future (Romm 2011). Realizing the difficulties, many countries are promoting the ancient concept of water harvesting. Like large dams, the smaller check dams can also hold large volume of water from seasonal rains solving local scarcity while improving socio-economic conditions of peasants. But, what are the genuine eco-economic benefits offered by the low-tech check dams? A closer look at the services rendered by them may lend a hand to grasp their long ignored potential serving nature, culture and humanity.

Irrigated Drylands

Big dams have expanded irrigated areas for decades; they also reduced famine and hunger worldwide. Yet, building more dams does not mean that the irrigated areas will automatically increase. For example, India's Maharashtra state holds the highest number of big dams (1,845). But, it has the lowest irrigated area of only 18%, which continues to push farmers in distress (Dhara, 2016). Furthermore, the distribution of irrigation water has often been restricted to the dam-linked canal system that depends on gravity to transfer the fluid commodity. Drylands are often out of the reach of big dams. Besides, the undulating topography of dryland makes it even less conducive for irrigated agriculture. Therefore, people are historically dependent on rain-fed farming; they use traditional methods using check dams and percolation tanks to store water. In fact, over 60% of India's crop area comes under rain-fed farming. Similarly, rain-fed farms contribute from 10 to 70% of the total GDP in most African countries (Biazin *et al*, 2012). Likewise, China depends on rains to sustain the drylands (Yuan *et al*, 2003).

Data from high resolution satellite show evidence of check dams in the arid Anantapur district of southern India stabilizing water levels in village wells leading to increase in irrigated area (Raghu and Reddy, 2011). Also, check-dams of Kerala in southern India provided water in summer months expanding irrigated area (Balooni *et al* 2008). Data from western India also support the perception of check dams significantly expanding irrigated area (Jagawat, 2005; Phansalkar and Verma 2005; Agoramoorthy, 2015). Furthermore, a recent report showed evidence of 385 check dams erected in western India as of 31 March 2016 expanding 57,926 acres of irrigated area, benefitting 24,885 families or 149,310 people (Jagawat, 2016).

Livestock and Wildlife Depend on Check Dams

India continues to lead the world in livestock population. A national survey done in 2012 showed that India holding over half billion livestock that include cattle, buffalo, sheep, goat, pig, horse, mule, donkey, camel and yak (GoI 2012). Most domestic animals tend to free-range in open fields cross India and they often rely on natural water holes to relive thirst. But, most of the shallow waterholes dry out in summer months. But, check dams alternatively provide ample of water to numeus livestock (Figure 1).

Figure 1. Livestock utilizing water from check dams on hot summer day in India's drylands (Photo courtesy Sadguru Foundation).

Several sanctuaries and national parks in India suffer from water shortages each summer due to extreme heat. About 150 panthers died in 2016 due to water scarcity that plagued sanctuaries and national parks. A study showed increasing conflict between humans and wildlife during summer months since animals were forced to go near villages in search of water (Sinu and Nagarajan, 2015). When check dams are built in forest areas, they provide enough water in summer months to large number of wildlife. For example, a check dam located near the sloth bear sanctuary at Ratnamahal in western India continues to provide drinking water to numerous wild animals during summer months (Agoramoorthy, 2015). Similarly, in southern India, 35 check dams were built by the forest department covering many rivers originating from the Western Ghats rainforest and they extended water to numerous species of wildlife, eventually reducing human-wildlife conflict (Viju, 2013). The above cases portray that check dams have the aptitude to decrease human-wildlife encounters during summer months in wild landscape.

Revival of Forest and Reduction of Erosion

Forest vegetation grows naturally along rivers, which is ecologically termed as the riparian zone. The forest growth along rivers organically restores soil and reduces erosion (Bren, 2016). Before the implementation of small dams, there was hardly any forest during summer months in the drylands of western India. But, satellite data analyzed over a decade showed evidence of dense forest vegetation along check dams in the arid state of Rajasthan since the backwater covered huge area (Figs. 2 and 4; Agoramoorthy and Hsu 2016). Likewise, Fu *et al* (2000) reported that between 1984 and 1996, China's overall forest cover increased by 36% while the slope farmland decreased by 43% after building check dams. Besides, the change in patterns of land usage reduced soil erosion nearly by means of 24% yearly (Fu *et al* 2000). Moreover, the check dams revived Mediterranean vegetation in Spain and Italy (Bombino *et al* 2006; Boix-Fayos *et al* 2007). They also increased forest growth in eastern India (Bhave and Raghuwanshi, 2014).

A report by Abedini et al (2012) shows check dams in Malaysia reducing the rate of soil erosion significantly. In addition, they acted as micro-percolation tanks to increase water filtration into subsurface, thereby retaining soil moisture used by vegetation for growth (Sakthivadivel, 2007). The expansion of forest vegetation along river banks also reduced soil erosion leading to sediment load reaching downstream (Matta, 2009). Similarly, check dams were created in the Alpine streams in Europe for over a century to control sediments. A review by Piton *et al* (2016) described details on the uniqueness of Alpine check dams with examples from France. Furthermore, the Japanese check dams in Furnao river catchment included 72 dams and they were reported to protect mountainous ecosystems (Chanson, 2004).

Retaining Carbon

The cultivated soils worldwide have lost enormous amount of their original carbon stock to contact with air. Scientists therefore are expanding investigations on how carbon sequestration takes place in soils so that the heavily depleted carbon could be restored in the formerly cultivated great plains of North America, Northern China and Australia. The check dams however do the same job effortlessly and at a fraction of the cost. A study by Lu *et al* (2012) for example estimated the carbon retention effects of medium to large-sized check dams in China's Loess Plateau and they retained over 42 million tons of soil organic carbon with high spatial variability. Hence, they concluded it to be a huge amount close to 1.5% of the soil organic carbon stored in soil layer across the entire plateau. Remarkably, the reported amount comes to 4% of the estimated total fossil fuel carbon emission from whole of China during the year 2000 (Lu *et al* 2012).

China by the way, has built numerous check dams without gates and they apparently alleviate climate change consequences through carbon cycling. But data are lacking for countries such as India, which continues to build numerous check dams. Therefore, more research on the role of dams in carbon cycling is needed with statistics on the exact number of working check dams from India and elsewhere.

Reduction in Sediment Load

Check dams are known to reduce the dispersal of coarse sediments. For instance, Ran *et al* (2008) have analyzed the granular sediment withholding in five major catchments of Hekou-Longmen midstream of China's enormous Yellow river, which is the 3rd longest in Asia and 6th longest globally. It transports high coarse sediment concentration. They compared the efficiency of check dams in reducing sediment loads with other soil conservation measures. They concluded that check dams efficiently and quickly reduce coarse sediment loads flowing into the river. They also recommended maintaining the percentage of check dams at 3% in the region to decrease coarse sediments draining into the river effectively (Ran *et al*, 2008).

Check dams are known as Sabo in Japanese that literally translate to sand protection. They reduce excess sediment load, thereby preventing river degradation while regulatory debris flow in rivers (Chanson 2004; Itoh et al 2013). A study conducted in China's Duozhao catchment of Jiangjia stream by Zeng et al (2009) evaluated the longitudinal performance of 44 check dams over 25 years. They showed significant reduction in soil erosion, from 2.7 to 0.3 m yr-1 and also sediment loading. Similarly, studies from the eastern Himalayan Mountains of India showed evidence that check dams reducing sediments over 50% into the Umroi watershed (Singh *et al*, 2011).

Groundwater Recharge

Check dams are known to rejuvenate dry landscape into a hydrated ecosystem since they can store enormous amount of water (Figure 2). So, even a simple and small check dam can serve as an excellent artificial recharge structure if it's built in monsoon-dependent rivers with an aim to store surface runoff caused by rainfall. Studies conducted on check dam reservoirs dominated by granites, basalts and sandstones show that they outstrip chronic droughts plaguing arid land.

Figure 2. A view of dried up river Hiren at Borbhatod village in Rajasthan, India (above) and the hydrological transformation with full of water (below) after check dam built in 2014 (photo courtesy Sadguru Foundation). This check dam supports four lift irrigation schemes and large number of potable pump sets used by farmers to irrigate 800 acres. There 13 such check dams are located on river Hiren to harvest rainwater (Photo courtesy Sadguru Foundatio).

Reports from India's drylands show numerous cases of how check dams enhance groundwater recharge in open wells in villages (Jagawat, 2005, 2016; Phansalkar, 2005; Agoramoorthy, 2015; Agoramoorthy and Hsu, 2016). Similarly, studies have shown that the effective way to preserve soil and freshwater in China's Loess Plateau was through the construction of check dams (Xin *et al* 2004; Xu *et al* 2004; Liu *et al* 2006). Likewise, a network of 107 check dams built in the semi-arid southeastern Spain efficiently recharged groundwater (Martín-Rosales *et al* 2007). Also, Raju *et al* (2006) evaluated the impact of check dams and indicated that groundwater levels increased by 1.8 m in post and pre-monsoon periods in the Swarnamukhi river basin of southern India.

Gujarat state has proactively installed check dams in several villages, especially in the Saurashtra and Kutch regions. After suffering great losses due to drought in the year 2000, the government, with the help of social workers and NGOs implemented many water harvesting projects (Shah 2010). Funds were amassed under the Sardar Patel Participatory Water Conservation Program (SPPWCP) through voluntary money, labor and materials contributions from 2000 to 2001. About 10,700 check dams were constructed across Saurashtra, Kutch, Ahmedabad and Sabarkantha regions costing USD 28 million, with 40% matching funds from SPPWCP and the remaining from the state (Sakthivadivelu 2007; Shah 2010). In 2002, an independent study (Shingi and Asopa, 2002) evaluated the performance of 102 check dams built under the SPPWCP in 96 villages and concluded that a net 300 MCM of water recharged in an average rainfall year. This led to

30% increase in agricultural yield, with each check dam recharging at least 7 wells and 32,000 m² of agricultural land in the vicinity of just one check dam. Furthermore, the study recommended not only maintaining partnership between government and public, but also expanding to larger scales across similar drylands elsewhere (Sakthivadivelu, 2007).

Reduction of Groundwater Toxicity

The fluoride levels beyond 1.5 ppm in groundwater persist in several regions across the Indian subcontinent, which in due course leads to fluorosis. The easiest way to mitigate the fluorosis crisis in affected areas is to naturally dilute fluoride concentration in groundwater. Studies have shown that check dams built upstream of high fluoride areas in southern India significantly reduced fluoride levels in groundwater (Bhagavan and Raghu 2005; Brindha *et al* 2016).

Check dams are known to naturally recharge groundwater, therefore, they have the inherent ability to dilute and neutralize various types of toxins naturally found in the earth system and artificially introduced by human activities. However, more scientific studies are needed on how check dams reduce the concentration of toxic chemicals in groundwater. Arsenic and fluoride are known to be the two major ground water pollutants affecting many rural areas of the Indian sub-continent. So, check dams may have the potential to dilute such pollutants that continue to affect the health of millions of people and livestock in the region (Chouhan and Flora, 2010).

Cost-effective Check Dams

Funds are fundamental for building water harvesting infrastructures. But the services that the check dams provide to nature, culture and humanity are enormous as shown in Figure 3. The largest check dam at Baneshwar Dham (Rajasthan) was built by Sadguru Foundation in partnership with the government and corporate sponsors. When it was opened in 2007, it had a price tag of USD 1 million. The check dam (length 367 m; height 7.3 m) expanded 7,000 acres of irrigated area ultimately benefitting 18,000 peopl.

There is a big dam sited upstream called Mahi-Bajaj Sagar dam (length 3109 m; height 74.5 m) opened by the government in 1985 with a price tag of USD 300 million. The backwater from the big dam affected 98 villages and displaced 6975 households or 40,000 people (Jayaraj, 2014). The big dam expanded irrigation to over 150,000 acres of land or over twenty times of the irrigated area of Baneshwar Dham check dam (Agoramoorthy, 2015). If twenty more of these smaller dams are added in the river, it would cost over USD 20 million, with similar irrigation prospective of the Mahi-Bajaj Sagar large dam. But, the gated check dams will neither displace people nor damage environment.

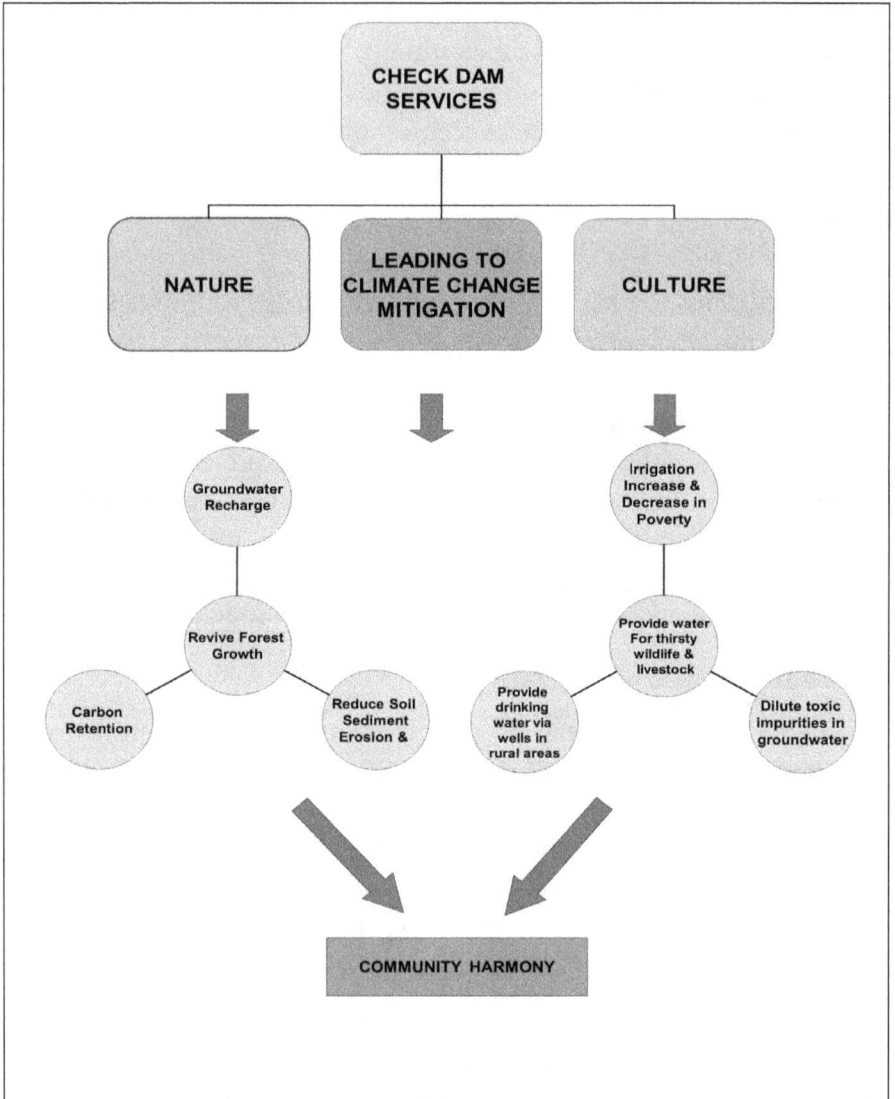

Figure 3. Graphical depiction of the services provided by eco-friendly check dams to nature, culture and humanity at large.

The cost to construct check dam is lower than that of other recharge infrastructures that produce same advantages. According to an economic analysis by Sakthivadivel (2007), in order for a single check dam to recharge 1000 m3 of groundwater, it would cost only USD 1. If one wants to reach the same amount of water recharge potential to reach similar benefits by using other technological options, it would cost more. For example, it would require USD 550 more to drill tube wells, USD 8 more to spread canal system, USD 515 more to expand recharge pits, and USD more 5 to include percolation tanks. So, the comparison of cost estimates clearly shows that other water harvesting methodologies are not cost-effective and eco-friendly. Above all, the check dams retain the environmental integrity while promoting community harmony at the grassroots across thousands of dryland villages (Figure 3)

Figure 4. A view of the check dam in Borekhedi village in Rajasthan, (above) built by Sadguru Foundation showing best quality gated check dam structure holding large volume of water (photo courtesy Sadguru Foundation); poorly designed and failed check dam structures (below) built by contractors without community support near Dahod (photo by S. Chaudhary).

Many countries have the tradition of building check dams for centuries. When engineers build check dams, they often block the natural water flow, which creates the problem of siltation leading to self-destruction of dams. Besides, the blockage of water flow in any river system obviously stops the movement of natural organisms (fauna, flora and fungi), minerals and organic materials that are crucial to maintain biodiversity in rivers and oceans. Therefore, it is absolutely necessary to build check dams with manually-workable steel-gates (Figure 2), supported by best quality materials such as bricks, stones, steel, cement and concrete. Good quality construction is crucial for check dams since they have to withstand the fury of the force of water. Gates should be opened during the monsoon season and only water from the last downpour at the end of monsoon should be harvested by closing gates.

Most check dams are frequently built by contractors without active participation of communities. As a result, the structures are ignored and the improper maintenance leads to malfunction (Figure 4). Thousands of such failed minor irrigation structures spread across many regions (Agoramoorthy, 2015). Therefore, active participation of local farming communities is crucial for the long-term management and preservation of check dams. It's about time for the society to realize the enormous potential of check dams to mitigate future climate change consequences.

Chapter 4

Impact of Irrigated Agriculture in Drylands

"The real cause of hunger is the powerlessness of the poor to gain access to the resources they need to feed themselves"—Frances Moore Lappe

Introduction

Asia has the largest stake of the arable land in the world (32%), which is followed by North America (17%) and Africa (14%). Besides, India leads the world in harboring the largest irrigated area with 96.4 million acres, which is followed by China with 46.9 million acres, and USA with 42 million acres, respectively (Renner, 2012). Agriculture productivity largely depends on irrigation water. So the irrigation sector consumes nearly 70% of all the freshwater withdrawals worldwide destabilizing the natural ground/surface water supply and demand cycle (Jarvis, 2014).

Irrigation expansion has shown a slowdown in India since the 1970s due to various problems including the decline in irrigation investment to poor performance of large canal irrigation systems combined with corruption and mismanagement in the infrastructure construction processes (Hussain and Hanjra 2004). India has also experienced rapid economic development often at the cost of natural environment (Bajpai, 2000). Ground water, which is crucial for agricultural development, has been severally depleted. Scientists argue that India's green revolution has gone brown due to the creation of agrarian class differentiation, poor soil quality, ecological degradation, decreasing yields and falling groundwater (Greenland 1997; Lipton and Longhurst 1989; Shiva 1992; Atkins 2001; Jha 2002; Kapila and Kapila 2002; Jagawat 2005; Sharma 2007; Agoramoorthy, 2009).

The downsides of India's irrigation and agriculture strategy are the historic neglect of catchment areas in remote drylands where tribal communities inhabit for centuries. Unlike the non-tribal communities that prefer to live in the plains and coastal areas, India's tribal societies often dwell in the semi-arid uplands and riverbanks (Phansalkar and Verma 2005). The commonly used flow irrigation or gravity irrigation in flatlands is not suitable for drylands due to rugged terrain where farms are at a higher level than the water source (25-40 m, Figure 1). Therefore, lifting water from its source to upland farms using mechanized pumps known as 'lift irrigation' is the best irrigation option for drylands (Jagawat 2005). The lift irrigation needs to be supported by water harvesting structures such as lakes, rivers, percolation tanks and check dams. In order to build the lift irrigation structures, funding is crucial hence partnership involving government, non-government and private corporations is necessary. Then only, rainwater could be harvested, stored and distributed to farms. In fact, past studies have shown that how irrigation could reduce rural poverty (Jagawat 2005; Chitale 1994; Brisco 1999; Bhattarai *et al* 2007).

Figure 1. Lift irrigation system established by Sadgur Foundation in Bhanasimal on Kadana reservoir in Panchmahal (Gujarat state) that provides irrigation supports to 650 acres (Photo courtesy Sadguru Foundation).

Although participatory approaches have developed in popularity over the last few decades to enhance positive change in marginalized communities, they are now increasingly used to deal with the complex issue of sustainable development (Bruges and Smith, 2008). Despite globalization, the governments' efforts to assist farmers in developing countries through agriculture extension and subsidies have not succeeded fully to meet the harsh realities of farming (Shiva, 1992). Therefore, developing countries have turned to participatory approaches to make agriculture more ecologically and economically sustainable (Pretty 1995; Singh and Ballabh 1996; Jagawat 2005). However, understanding the potentiality of sustainable agriculture requires thorough analysis of the nature, purpose, problem and prospect of community participatory approaches (Bruges and Smith, 2008).

Lift Irrigation Operation

Each lift irrigation unit includes components such as pump house to keep machinery, distribution system with underground pipelines, main delivery chamber located at the highest point where water is lifted, and electrical accessories (Figure 2). The transformer and power supply are provided by the government-run electricity board. The underground pipelines form a network in which water flows by gravity and are connected to distribution chambers. Villages located in Gujarat, Rajasthan and Madhya Pradesh States were visited to collect data on lift irrigation systems in check dams, reservoirs, rivers, canals, and tanks, and their irrigation benefits following methods described by Creswell (1994). A total of 391 lift irrigation systems were completed between April 1976 and March 2014 of which 382 are included in analysis.

Figure 2. Community lift irrigation scheme at village Borkhedi Mahudi falia Tehsil: Kushalgadh District :Banswra implemented by Sadguru Foundation under Rsshtriya Krishi Vikas Yojana ,Government of Rajasthan (Photo courtesy Sadguru Foundation).

Data on the year of operation, water source, motor pumping capacity, lifting heights, total cost, expansion of irrigated areas and number of beneficiaries were pooled from the archives of Sadguru Foundation. Discussions and interviews of 330 people were conducted while visiting lift irrigation corporatives in villages to record the impact in terms of increased irrigation area, improved agricultural productivity, and enhanced livelihood, following the methods of Mikkelsen (1995).

Community-irrigation Benefits

The community lift irrigation is managed by the user farmers in villages. The cost to construct lift irrigation units and check dams was sponsored by government agencies responsible for rural development through Sadguru Foundation with matching funds from private corporations. Tribal people who inhabit India's drylands own lands with an average holding of 2 acres, which gives them farming opportunity (Jagawat, 2005). The objectives of creating lift irrigation cooperatives in villages were to increase capacity building of farmers to enhance self-sufficiency in food production (Figure 3). Emphasis was given on community-oriented collective management of water and natural resources to promote sustainable livelihood and financial self-reliance. Peoples' involvement was ensured from the beginning and it began with meetings of farmers and discussions to raise awareness on issues surrounding collective management of irrigation oppotunities.

Figure 2. A farmer who benefitted enormously from lift irrigation stands proudly among the wheat crop in village Ambapada (Banswara district, Rajasthan). By getting access to irrigation water, the farmer got bumper crop not seen in the past (Photo courtesy Sadguru Foundation).

Once a farmer became a member of the community lift irrigation cooperative, he/she was eligible to get irrigation water from the participatory system. Irrigation cooperatives are generally managed by a committee of 12 elected members. The committee is headed by a chairman and assisted by a secretary, who keeps financial records and organize monthly meetings to inform members of activities. Apart from holding meetings, the committee oversees auditing, financial management, water distribution, collection of water charges, payment of electricity bills and staff salaries, maintaining or repairing lift irrigation systems, and solving water distribution disputes.

Figure 3. The Sadguru Foundation has trained women farmers to grow flowers, mangoes and Indian gooseberry or amla in villages of Rajasthan using water from lift irrigation and check dam (Photo courtesy Sadguru Foundation).

When some farmers are unable to pay their dues, they are allowed to pay in full before the next irrigation season with an annual interest of 20%. Then only the farmer can be entitled to receive water. All cooperatives save money from various agricultural activities (Figure 3) and profits are deposited in bank, which shows the self-sufficiency in irrigation. It also shows how farmers managed village-level institutions with efficiency and self-reliance. People lived in traditional houses built with mud and thatched roof prior to the implementation of lift irrigation schemes in villages.

But, afterwards the percentage of traditional houses in villages decreased dramatically from 94% (n=125) to 51%. The average number of traditional houses in six villages was 20 (± 8.6) and it decreased to 10 (±4.1). On the other hand, the percentage of semi-standard houses increased nearly nine times (4-35%) while the strongest standard houses also increased 8.5 times (2-14%). This indicates the economic benefits obtained from lift irrigation schemes contributing to the improvement in housing standards in villages. All households in the above villages faced food shortages each year due to droughts and lack of irrigation water. So people relied on migration to nearby towns in search of jobs. After the implementation of lift irrigation schemes, all households attained self-sufficiency in food grain production that lasted for the whole year. Hence the seasonal migration of farmers in search of jobs in nearby towns had stopped. It was due to increase in irrigated area to produce crops that were not possible before.

Problems Facing Lift Irrigation Schemes

The community lift irrigation schemes sometimes encounter difficulties involving theft of machinery, inability to obtain lost machinery, disputes, political interferences, inability to pay electricity bill, droughts, shortage of electricity, and lack of crop insurance or subsidy to farmers who face crop failure or natural disasters. The situation in Gujarat and Madhya Pradesh is serious due to the lack of government support to farmers who suffer crop losses during droughts. However, Rajasthan has been sympathetic to farmers and subsidizes or cancels electricity bill during droughts. This approach must be extended to other states.

Flood is one of the major natural calamities that affect the lift irrigation infrastructure and machineries. For example, flooding was serious during the 2012 monsoon when the rivers namely Hiren (Rajasthan), Hadaf and Anas (Gujarat) were swelled above shore damaging the lift irrigation systems. Then the community leaders requested the Sadguru Foundation to assess damage. Experts were sent to collect data and to work out repair strategy. The analysis was done within a week that pointed damages to civil structures (pump house and well), mechanical structures (pipes-succession, delivery, valves), and electrical structures (pump, motor, starter, panel board). After a meeting, the leaders of lift irrigation federations decided to cover 25% of the repair costs. Within a month, all 72 damaged lift irrigation systems were repaired and crops were irrigated without interruption. In the meantime, Sir Dorabji Tata Trust (SDTT) had allocated 80% of funds while the rest of the repair expenditure was covered by the farmers' irrigation federations. This shows the efficiently of village-level cooperatives in tackling the crisis timely. But the government-managed lift irrigation systems need to wait for long time since it is up to the courtesy of officials to respond to disasters. If houses are damaged or livestock killed during natural disasters, the government can pay for the loss rather quickly. But repairing community irrigation structures is not all that imperative from the government point of view to deal the crisis quickly. That's why most government-built lift irrigation schemes seldom function (Choudhry *et al* 2002).

Farmer Suicide and Community Irrigation

About 135,445 cases of suicide in 2012 were recorded across India (NCRB 2014). The self-employed category accounted for 38.7% of victims, of which 11.4% were engaged in farming. Age-wise profile of victims shows that nearly 36.7% of them were farmers (30-44 years of age). India has recorded over a quarter million farmers committing suicide between 1995 and 2010. In 2010 alone, 15,964 farmers committed suicide and it brings the cumulative 16-year total from 1995 when data collection started to 256,913, which is the worst-ever recorded farmer suicide in the history (NCRB, 2014). Due to India's recent economic outbursts, the cost of formers' basic needs has gone up while earnings are sinking due to the increase in price of seeds, pesticides and fertilizers. When farmers are faced with credit squeeze by legitimate banks, they are forced to borrow money from illegal loan sharks. After they are engulfed by debt due to repeated crop failures, suicide becomes the only salvage. But the government is aware of the fact that debt-ridden poverty is driving farmers towards suicide. Nevertheless, it has not come up with a workable strategy to provide sufficient emergency support funds to cover the cost of crop production ahead of the worst case scenarios involving crop failures (Gaiha 2000; Agoramoorthy 2008).

Although the crisis seems to be unmanageable, it can be tackled if farmers form community-based approach to deal irrigation shortages as shown in this paper. When farmers face financial hardships, the final safety net constitutes their friends, family and community. The formation of irrigation cooperatives can serve as 'social capital', which can be used by the families, friends, and associates of the impoverished farmers at times of crisis. When communities form social networks similar to the irrigation cooperatives discussed in this paper, they are in a comfortable position to confront poverty, escape from suicide (North 1990; Narayan 1995), resolve social disputes (Varshney, 2000), and ultimately take advantage of rural development economic opportunities (Isham 2000). Therefore, poverty and suicide can be naturally neutralized by reviving participatory approaches at the grassroots in villages.

As a matter of fact, the participatory approach is not new for India, and it was first introduced in 1904 when the Cooperative Credit Societies Act was ratified. Later the act was amended in 1912 to include non-credit institutions and federal organization (Singh and Ballabh, 1996). Besides, the government enacted the Multi-state Cooperative Societies Act in 2002 to provide democratic and autonomous working of cooperatives. The village-level irrigation cooperatives highlighted in this paper are united by state-level federations and registered under the cooperative societies act. When people collaborate to create their own social rules, opportunities for individuals and collective empowerment can emerge (Ostrom 1992; Singh et al 2001). Participatory approaches to manage natural resources including irrigation, forestry, salt mining, watershed, and fisheries were known to not only improve livelihoods, but also minimize suicides (Singh and Ballabh 1996; Beck 2001).

The success of lift irrigation systems built by the Sadguru Foundation was due to following reasons: i) use of high quality construction materials, ii) best design by experienced civil engineers, and iii) strong community mobilization combined with capacity building by experienced social workers (Agoramoorthy 2009). Unfortunately, government agencies while building minor irrigation infrastructures do not emphasize participatory approaches. Without the participation of communities, even highly funded development projects are doomed to fail or short-lived (Jagawat, 2005). For example, in Jhabua District of Madhya Pradesh, the government has installed over 1000 lift irrigation systems during the last decade, but 70% of them have failed (Choudhry *et al* 2002). Similarly, majority (80%) of the check dams built across China's Shanbei region between 1977 and 1978 has failed due to poor construction, flawed site selection, and lack of participatory approach (Xu *et al* 2004).

The financing of water-related infrastructure for decades was heavily dependent on government funds. The global infrastructure financing for telecommunications, power, transport and water accounted nearly half of all government spending. But the results had been perceived as unsatisfactory (Brisco, 1999). In the last decade, about 15% of the infrastructure investment in developing countries came from the private sector. Therefore there is a need for financial partnership involving government and private corporations with NGOs as catalysts to promote sustainable development. For example, along the US and Mexico border, NGOs serve as catalyst for public-private water infrastructure (Lemos *et al.* 2001). In Jordon, a successful public-private partnership has been reported in the management of domestic water sector using metaphors from ecology (Al-Jayyousi, 2003). The tri-sector partnership is therefore essential to minimize poverty at the grassroots across developing countries.

Sustainable Agriculture

Agricultural system is considered sustainable only when productivity is maintained over a long period by enhancing conservation of natural resources with significant profitability to guarantee financial benefit for farmers (Kessler, 1994). Since agricultural production is linked to ecology, consideration of interactions between agriculture and ecosystems has become a basic requirement to evaluate sustainability (van Wiren-Lehr 2001). India needs to double the food grain productivity by 2020 for domestic consumption. So, it is crucial to reduce overshoot of bio-capacity. Five major factors that determine the extent of global overshoot on bio-capacity; they include population, consumption of goods/services per person, resource use intensity, bio-productive areas, and bio-productivity per acre (WWF, 2006). An easier approach to reduce the overshoot is to transform drylands by building adequate water harvesting and distribution structures (Shah et al 1998, Jagawat 2005; Agoramoorthy, 2009). The fact of agriculture being an essential sector in rural development can neither be ignored nor be denied (Jewitt and Baker, 2007). Drylands must be focused to increase food productivity if India aims to succeed in sustainability, without creating negative consequences to nature.

Furthermore, human population pressure has been blamed for a series of ecological calamities in the Indian sub-continent where the density is 236 people per km^2, more than a thousand times of the Amazon region (Agoramoorthy and Hsu, 2002). Although population pressure, water shortages and desertification may increase in the region, it can be averted by change in attitude that involves mobilizing community support to integrate water resources with agricultural development, and to focus more on sustainable agriculture initiatives (Gleick, 1999). The best way to integrate water resources management in drylands is by controlling rainwater run-offs by constructing series of check dams in rivers (Agoramoorthy and Hsu, 2013).

India's economy is the fourth largest in the world in terms of purchasing power parity with a GDP of US$ 4.042 trillion. Despite its impressive growth, the dividends of growth failed to trickle down to the impoverished farmers. In order to eradicate poverty, India must strengthen rural development by integrating people with government, non-government and corporate support. Several NGOs have integrated the above on a smaller scale (Phansalkar and Verma, 2005; Shah *et al* 1998; Minj 1999; Kashwan 2006). A common perception in India is that NGOs complain more, but deliver less; they hinder economic progress.

Some have been implicated in misusing funds while others did excellent work, especially in water resource management, education, organic farming, and health (Anand and Anand, 2007). But they are often small, restricted to particular region to assist only limited communities and still struggling to find models that can be successfully implemented across India's diverse regions. So, it is imperative for NGOs that are involved in rural development to have transparency, accountability and responsible utilization of public funds. Besides, social responsibility should not be limited to large corporations and greater participation from small businesses and NGOs are absolutely necessary. Even small NGOs can significantly contribute for socioeconomic and political changes at the grassroots level. For example, the Sadguru Foundation's programs have given employment opportunities to over 1.6 million farmers through various farming activities.

India and China are among the top 10 nations most vulnerable for food shocks since they harbor large human populations. Food is the largest single component of household spending, nearly 80%, in countries with low per capita incomes, compared with 15% of the average family in EU or USA (Starke, 2008). Sustainable development in the agriculture sector may therefore hold the key for the future survival of humanity. The revival of the community irrigation cooperative model at the grassroots pioneered by the Jagawats has the potential to improve agricultural output, minimize local food insecurity, reduce farmers' suicide, preserve water resources, and ultimately lessen rural poverty.

Chapter 5

Need for More Rural Development in Rajasthan

"Poverty is the worst form of violence"—Mahatma Gandhi

Introduction

Majority of the almost seventy percent of the people in India live in rural areas and statistics show that the maximum number of over 15 crores is concentrated in rural Uttar Pradesh while Mumbai leads the urban areas with over 5 crores. Some rural areas are too remote for the government's support system to reach. An example is the far-off ravines of Chambal Valley. For centuries, the Chambal valley has been the homeland to the feared bandits; many of whom liked to take away from the rich to be given to the poor in a kind of Robin Hood style episode that carefully cultivated good will gesture for generations in the wilderness. Arable land is obviously precious in Chambal drylands, and it is not difficult to see disputes over land ownership. My journey through the Chambal valley few years ago to study water resources management projects implemented by Sadguru Foundation actually revealed some secrets of the local communities, who have been living on the societal edge based at the rugged terrain dominated by gorges.

Smoke along the River

Accompanied by Sathish Mishra, a field staff who works for Sadguru Foundation in Rajasthan, I hiked along of one of Chambal's tributary, river Kali Sindh. After reaching Bor Khedi village that comes under Jhalawar district, I saw smoke bellowing near the water's edge. A meter long crocodile was basking and ignoring the frenzy on the opposite bank that reminded me that I was walking through the remote wilderness of Rajasthan. When I asked Mishra about the smoke, he whispered, "Kanjars making country liquor" and told me to stay quiet. He then walked for a mile, spoke to the people on the river bank, and got their permission for a personal chat. I learnt later that people usually brew illicit country liquor along the secluded river banks and ravines to avoid poice raids.

Figure 1. Check dam located on Chambal's tributary, the river Kali Sindh in Rajasthan state (above) and children play along the river while one pays attention to brew the illicit country liquor while parents engage in farm work (Photo by G. Agoramoorthy).

Joyful Children in Wilderness

I was surprised to see over a dozen joyful children, from toddlers to teenagers, burning firewood to prepare the country liquor under the watchful eyes of eight adults; I counted five men and three women in total. While photographing the smoke-filled liquor pots, I was about to stumble on to open cables that were lying on the ground. Mishra cautioned me not to step forward on the obvious death traps!

The cables transport stolen electricity from the nearby power line to private motors of farmers who were brewing the liquor to irrigate their farms. The adults were vigilant in monitoring the movements of strangers and also pumping water from the check dam build by Sadguru Foundation (Figure 1) to irrigate their crops. The children were happily playing along the riverbank and also helping their family members with the work. They had no idea that they were part of being co-conspirators to their parents' illegal and dangerous endeavors of the river bank.

Underprivileged Lifestyle

I asked an elderly woman why the children were working there and not going to school. Munna Bhai, a mother of three with an elegant smile answered that her community no longer involve in looting and killing. She was happy to say that water stored in the check dam help them a lot to irrigate their fields, using the stolen electricity of course, plus the supplement income derived from the illicit liquor business that provides them enough money to sustain their lives in the developmentally backward region.

I asked any possibility in future to stop the risky occupation of stealing electricity and producing illicit liquor, and she replied that if the government could assists them in establishing a lift irrigation system, they could irrigate more lands and then there would be no need to steal electricity or indulge in illicit liquor. She also said that her village is remote so it's naturally cut off from civilization. The faraway government offices have not at all helped the small village community for nearly over a century. All the adults unanimously said that the lift irrigation will improve their livelihood, which would in turn put their children back in school. The village head said that the stolen electricity is available only for five hours daily and about 15 households irrigate 10 to 15 acres of dryland.

Need for Rural Development Efforts

Stealing electricity out of necessity occurs not only in the remote Chambal valley, but also in the slums of New Delhi metropolis where power theft is almost a way of life for the slum dwellers. The method is simple, which involves an overhead power cable attached to a metal hook over it, and then a wire runs from the hook to the house or motor wherever the power is needed. I was traumatized though to hear the dangers involving the open cables transporting stolen electricity.

Within two years, three people, aged 5, 6 and 40 and six animals that include goats, cattle and buffalo were electrocuted by accidentally stepping on the open cables. Later, an irrigation engineer from Sadguru Foundation told me that one lift irrigation system can easily be set up at Bor Khedi with a cost of 20 lakhs if the Government is keen to help the community. One such system could irrigate about 50 acres. So, there is enormous need for small-scale rural development work in Rajasthan that can start with basic watershed management.

Figure 2. Once proper watershed development activities influenced by percolation tanks and check dams are constructed in drylands, ground water level naturally increased in numerous village wells in Banswara district of Rajasthan (Photo courtesy Sadguru Foundation).

The drylands have huge opportunities to build percolation tanks to harvest rainwater in addition to check dams (figure 2). As a matter of fact, Sadguru Foundation has built numerous large percolation tanks and check dams in Rajasthan and those played a major role to increase the ground water levels naturally (Figure 2). Some of the large percolation tanks surprisingly transformed the arid landscape. Later, I spoke to Harnath Jagawat who said that rural development work in India is a never ending job and it will go on until all villages across the country are uplifted from poverty.

During the British occupation of India, the Kanjar community was considered to be a criminal tribe according to the provisions of the Criminal Tribes Act of 1871. India's first Prime Minister, Jawaharlal Nehru opposed this act vehemently and he said in 1936 that the monstrous act should be removed. Naming any human tribe as criminal in nature only shows the discriminatory attitude of the ruling colonial regime of the past. Five years after India got independence from Britain, the term "criminal tribe" was removed from the statute book in 1952. But, social workers argue that tribal communities such as the Kanjars, Sansis, and Pardhis continue to be discriminated socially and by law enforcement authorities till today, despite the fact that many try to make a decent living through agriculture.

Chapter 6

Concluding Remarks

"We forget that the water cycle and the life cycle are one"— *Jacques Cousteau*

Over the last decade, I have interviewed hundreds of villagers in Rajasthan, Gujarat and Madhya Pradesh states where Sadguru Foundation has been working for many years. All admitted that ever since Sadguru Foundation built the lift irrigation systems and check dams, the crop productivity in villages increased and people attained self-sufficiency in food production. As a result, livelihoods of farmers improved leading to stopping of migration to cities in search of work. Moreover, villages got access to electricity, roads, schools, hospitals, *etc*. Now, village communities are more eager and ambitious to approach government agencies without hesitation for rural development projects.

While visiting a small village called Khejaria in Rajasthan where the river Chambal runs wild, people told me that some government officials collected funds for a cooperative lift irrigation. Sadly, the scheme never took off. Therefore, village elders approached Sadguru Foundation with a request and the Jagawats helped them to establish a new lift irrigation system. Now, farmers are happy with their cooperative society and looking for more options to start other types of agricultural business activities. These examples show that how the Sadguru Foundation has created not only sustainability in food production and elimination of poverty, but also cultured business-oriented approach on the minds of many still uneducated village folks. So, the Jagawats have created a unique cultural heritage at the grassroots with the philosophy of hard work leading to livelihood improvement.

While visiting the Sadguru Foundation's field office in Rajasthan, the engineer showed me a large file of letters of request to establish new lift irrigation systems and check dams from several villagers and some even begged for help. The situation to establish minor irrigation infrastructures in villages appear to be desperate. It also shows that there will be no end to rural development work in drylands. A new approach may be needed to build more check dams and lift irrigation systems by involving and training other NGOs in the region, which would relieve the existing pressure in Sadguru's field office in Rajasthan.

Working with the impoverished rural communities is not an easy job and in fact, convincing them to adopt new agricultural methods and practices is even more difficult. For example, when the Jagawats started the biogas project in villages, not many people were interested. Then, they had to support two families the entire cost of setting up the household biogas systems. Then the model was shown to neighbors and afterwards, one by one started to approach the Jagawats for help. In general, biogas is used by the developed non-tribals in urban areas and tribal people tend to use firewood for cooking. So, it created a kind of myth among the tribals that biogas fits only for the privileged urbanites and not others. This mind set was slowly changed due to the Jagawats' hands-on approach working with communities that live at the bottom of social hierarchy. This sort of practical approach is not taught at any university lecture halls so I consider the Jagawats as the best teachers in the field of rural development.

While visiting a village called Sankarpura, I met a gentleman named Sowsinghbhai, who was a community leader. He told me that before the Jagawats launched the lift irrigation project in his village, only 500 people (100 households) lived there, all in mud-houses with zero literacy level. He added that about 30 years later, over one hundred children go to school, twelve teachers came out of the village, and the population increased to 3500 with over one hundred university graduates. I saw one graduate walking with a shovel, ready to do farm work. Each household has 4 to 10 buffaloes or cows and they continue to grow crops using the lift irrigation and also grow eucalyptus trees. He said that the village was able to reach high economic sustainability only with the help of the Jagawats, and without them, we would still suffer in poverty.

Most of India's seventy million tribal people are impoverished; they constantly face survival challenges. The foremost among them are the marginal conditions to improve agriculture influenced by unreliable water supply. Therefore, focus has to be on drylands if India needs to succeed in agriculture sustainability, without creating negative consequences to natural environment. India's agriculture sector largely depends on water so it is more vulnerable for future climate change impacts. Studies have predicted that India may experience warmer and wetter weather conditions in future due to global warming and the summer monsoon may become more severe.

The water-centered rural development pioneered by the Jagawats has enormous potential to protect the integrity and functioning of river basins and aquifers in India's drylands. This is vital, as rivers are crucial for the continuation of human survival. It is therefore essential for the government to support building of more check dams across rivers to complement large dams in partnership with private corporations and non-government organizations.

I asked Harnath how many check dams are needed in India's drylands and he instantly replied that a minimum of 50,000 is urgently needed. Therefore, the low-tech option of building more checks dams linked by lift irrigation systems are in fact simple, environment-friendly and inexpensive when compared to other irrigation technologies. If the Sadguru model of rural development is adopted across India's drylands and elsewhere, it has great potential to increase agricultural output; guarantee food security; enhance groundwater; and, above all, mitigate the dangers predicted by the brutal climate change scenarios.

Literature Cited

Abedini, M., Said, M.A., & Ahmad, F. (2012). Effectiveness of check dam to control soil erosion in a tropical catchment. *Catena* 97, 63-70.

Agoramoorthy, G. 2008. Can India meet the increasing food demand by 2020? *Futures* 40:503-506.

Agoramoorthy, G. 2009. Sustainable development: The power of water to ease poverty and enhance ecology. Delhi: Daya publishing house.

Agoramoorthy, G. 2015. Sadguru model of rural development mitigates climate change in India's drylands. Daya Publishing House, Delhi.

Agoramoorthy, G., and M. J. Hsu, 2002.Threat of human-induced climate change cannot be ignored. *Current Science* 82:904-905.

Agoramoorthy, G., and M. J. Hsu, 2013. Partnership for poverty reduction in Gujarat, India: A case study of Sadguru foundation's water resource development work. *Asia Pacific Journal of Social Work and Development* 23:59-70.

Agoramoorthy, G., & Hsu, M.J. 2016. Small dams revive dry rivers and mitigate local climate change in India's drylands. *International Journal of Climate Change Strategies and Management* 8, 271-285.

Albinia, A. 2010. Empires of the Indus: The story of a river. W. W. Norton & Company; New York.

Al-Jayyousi, O. R. 2003. Scenarios for public-private partnerships for water management. A case study from Jordon. *International Journal of Water Resources Development* 19:185-201.

Anand, R., and U. Anand, 2007. India needs its NGO. Harvard International Review, 8 January (hir.harvard.edu).

Atkins, P., and I. Bowler, 2001. Food in society: Economy, culture and geography. London: Arnold.

Balooni, K., Kalro, A.H., & Ambili, G.K. 2008. Community initiatives in building and managing temporary check-dams across seasonal streams for water harvesting in South India. *Agricultural Water Management* 95, 1314-1322.

Beach, T., Luzzadder-Beach, S., Dunning, N., Hageman, J., & Lohse, J. 2002. Upland agriculture in the Maya lowlands: Ancient Maya soil conservation in Northwestern Belize. *Geographical Review* 92, 372-397.

Beck, T. 2001. Building on poor people's capacities: The case of common property resources in India and West Africa. *World Development* 29:119-133.

Bhagavan, S.V., & Raghu, V. 2005. Utility of check dams in dilution of fluoride concentration in ground water and the resultant analysis of blood serum and urine of villagers, Anantapur district, Andhra Pradesh, India. *Environmental Geochemistry and Health* 27, 97-108.

Bhattarai, M., R. Barker, and A. Narayanamoorthy, 2007. Who benefits from irrigation development in India? Implications of irrigation multipliers for irrigation financing. *Irrigation and Drainage* 56:207-225.

Bhave, A.G., Mishra, A., & Raghuwanshi, N.S. 2014. A combined bottom-up and top-down approach for assessment of climate change adaptation options. *Journal of Hydrology* 518, 150-161.

Biazin, B., Sterk, G.; Temesgen, M., Abdulkedir, A., & Stroosnijder, L. 2012. Rainwater harvesting and management in rainfed agricultural systems in sub-Saharan Africa: *A review. Physics and Chemistry of the Earth, Parts A/B/C*, 47-48, 139-151.

Boix-Fayos, C., Barbera, G.G., Lopez-Bermudez, F., & Castillo, V.M. 2007. Effects of check dams, reforestation and land-use changes on river channel morphology: Case study of the *Rogativa catchment* (Murcia, Spain). *Geomorphology* 91,103-123.

Bombino, G., Tamburino1, V., Zimbone, S.M. 2007. Assessment of the effects of check-dams on riparian vegetation in the Mediterranean environment: A methodological approach and example application. *Ecological engineering* 27,134-144.

Bren, L. 2016. Forest hydrology and catchment management: An Australian perspective. Springer, Berlin.

Brindha, K., Jagadeshan, G., & Kalpana, L. 2016. Fluoride in weathered rock aquifers of southern India: Managed aquifer recharge for mitigation. *Environmental Science and Pollution Research* 23, 8302-8316.

Brisco, J. 1999. The changing face of water infrastructure financing in developing countries. *International Journal of Water resources Development* 15:301-308.

Bruges, M., and W. Smith, 2008. Participatory approaches for sustainable agriculture: a contradiction in terms? Agriculture and Human Values 25:13-23.

Cao, S., Liu, X., & Huang, E. 2010. Dujiangyan irrigation system: A world cultural heritage corresponding to concepts of modern hydraulic science. *Journal of Hydro-environment Research* 4, 3-13.

Chanson, H. 2004. Sabo check dams-mountain protection systems in Japan. *International Journal of River Basin Management* 2, 4:301-307.

Chitale, M. A. 1994. Irrigation for poverty alleviation. *Water Resources Development* 10:383-391.

Chouhan, S., & Flora, S.J.S. 2010. Arsenic and fluoride: Two major ground water pollutants. *Indian Journal of Experimental Biology* 48, 666-678.

Choudhry, K., H. Shah, and H. Jagawat 2002. A Study of Government-installed Lift Irrigation Schemes in District Jhabua, Madhya Pradesh. Dahod, India: Sadguru Foundation.

Creswell, J. W., 1994. Research design: Qualitative and quantitative approaches. Los Angeles: Sage.

Dhara, T. 2016. Marathwada drought: Maha has the most dams in the country, but the least effective irrigation network, leaving lakhs in the lurch. www.firstpost.com/india/marathwada-drought-maha-has-the-most-dams.

Ehrlich, P.R., Holdren, J.P., and Ehrlich, A.H. 1978. Eco-science: Population, resources, environment. W.H. Freeman & Co., Gordonsville, VA.

Fu, B.J., Chen, L.D., Ma, K.M., Zhou, H.F., & Wang, J. 2000. The relationships between land use and soil conditions in the hilly area of the loess plateau in northern Shaanxi, China. *Catena* 39, 69-78.

Gaiha, R. 2000. Do anti-poverty programs reach the rural poor in India? Oxford Development Studies 28:71-95.

Gleick, P. H. 1999. The human right to water. *Water Policy* 1:487–503.

GoI. 2012. Government of India's 19th livestock census 2012. Department of Animal Husbandry, Dairying and Fisheries, Ministry of Agriculture, Delhi.

Greenland, D. J. 1997. The sustainability of rice farming. Wallingford: CAB International IRRI.

Hussain, I., M. Hanjra, 2004. Irrigation and poverty alleviation: Review of the empirical evidence. *Irrigation and Drainage* 53:1-15.

Isham, J. 2000. The effect of social capital on technology adoption: Evidence from rural Tanzania. IRIS Center Working Paper 235. Maryland: University of Maryland.

Itoh, T., Horiuchi, S., Mizuyama, T., & Kaitsuka, K. 2013. Hydraulic model tests for evaluating sediment control function with a grid-type Sabo dam in mountainous torrents. *International Journal of Sediment Research* 28, 511-522.

Jagawat, H. 2005. Transforming the dry lands. The Sadguru story of western India. India Research Press, Delhi.

Jagawat, H. 2011. Pictorial document of green revolution by marginal farmers through water centered NRM program in semi-arid tribal drylands. A report published by Sadguru foundation, Dahod.

Jagawat, H. 2012. Brief note on proposed revitalizing rainfed agriculture. A report published by Sadguru foundation, Dahod.

Jagawat, H. 2016. Annual report on the progress under various programs cumulative physical achievements till 31 March 2016. Sadguru Foundation publication, Dahod, India.

Janaiah, A., M. L. Bose, A. G. Agarwal, 2000. Poverty and income distribution in rainfed and irrigated ecosystems: Village studies in Chattisgarh. Economic and Political Weekly 30:4664-4669.

Jayaraj, S.S. 2014. Dam induced development- a study of Mahi Bajaj Sagar project in Banswara, Rajasthan (India). *International Journal of Development Research* 4, 844-851.

Jarvis, W.T. 2014. Contesting hidden waters: Conflict resolution for ground water and aquifers. Earthscan, London.

Jewitt, S., and K. Baker, 2007. The green revolution re-assessed: Insider perspective on agrarian change in Bulandshahr District, western Uttar Pradesh, India. Geoforum 38:73-89.

Jha, P. 2002. Land reforms in India. Delhi: Sage.

Kapila, K., and U. Kapila, 2002. Indian agriculture in the changing environment. Delhi: Academic Foundation.

Kashwan, P. 2006. Traditional water harvesting structure: community behind community. *Economic and Political Weekly* 18:596-598.

Kessler, J. J. 1994. Usefulness of human carrying capacity concept in assessing ecological sustainability of land-use in semi-arid regions. Agriculture Ecosystem and Environment 48:273–284.

Lemos, M. C., D. Austin, and R. Merideth. 2001. Public-private partnerships as catalyst for community-based water infrastructure development: the boarder water works program in Texas and New Mexico. *Environment and Planning Government and Policy* 20: 281-295.

Lipton, M., and R. Longhurst, 1989. New seeds and poor people. London: Unwin and Hyman.

Liu, Y.W., Yen, T., et al 2006. Erosive resistibility of low cement high performance concrete. Construction and Building Materials 20, 128–33.

Lu, Y., Sun, R., Fu, B., & Wang, Y. 2012. Carbon retention by check dams: Regional scale estimation. *Ecological Engineering* 44, 139-146.

Luzzadder-Beach, S., Beach, T.P., & Dunning, N.P. 2012. Wetland fields as mirrors of drought and the Maya abandonment. *Proceedings of the National Academy of Sciences of the United States of America* 109, 3646-3651.

Mccorriston, J., & Oches, E. 2001. Two early Holocene check dams from Southern Arabia. *Antiquity* 75, 675-676.

Matta, J.R. 2009. Rebuilding rural India: potential for further investments in forestry and green jobs. *Unasylva* 233: 36-41.

Mikkelsen, B. 1995. Methods for Development Work and Research: A Guide for Practitioners. Delhi: Sage.

Minj, S. 19999. Lift irrigation to lift from poverty. Delhi: Manak Publication.

Molden, D. 2007. Water for food, water for life. Earthscan, London.

Martín-Rosales, W., Gisbert, J., et al 2007. Estimating groundwater recharge induced by engineering systems in a semiarid area (southeastern Spain). *Environmental Geology* 52, 985-995.

Narayan, D. 1995. Designing community-based development. Washington, D.C.: World Bank.

NCRB, 2014. Accidental death and suicide in India. Delhi: National Crime Records Bureau, Ministry of Home Affairs.

North, D. 1990. Institutions, institutional change, and economic performance. Cambridge: Cambridge University Press.

Ostrom, E. 1992. Crafting institutions: Self-governing irrigation systems. San Francisco: Institute for Contemporary Studies.

Phansalkar, S., & Verma, S. 2005. Mainstreaming the margins: Water-centric livelihood strategies for revitalizing tribal agriculture in central India. Angus & Grapher, Delhi.

Piton, G., Carladous, S., Recking, A., et al 2016. Why do we build check dams in Alpine streams? An historical perspective from the French experience. Earth Surface Processes and Landform, doi:10.1002/esp.3967.

Pretty, J. 1995. Participatory learning for sustainable agriculture. World Development 23:1247-1263.

Raghu, V., & Reddy, K.M. 2011. Hydro-geomorphological mapping at village level using high resolution satellite data and impact analysis of check dams in part of Akuledu Vanka watershed, Anantapur District, Andhra Pradesh. *Journal of Indian Geophysical Union* 15, 1-8.

Raju, N.J., Reddy, T.V.K., & Munirathnam, P. 2006. Subsurface dams to harvest rainwater—a case study of the Swarnamukhi River basin, Southern India. *Hydrogeology Journal* 14, 526–531.

Ran, D.C., Luo, Q.H., Zhou, Z.H., Wang, G.Q., & Zhang, X.H. 2008. Sediment retention by check dams in the Hekouzhen-Longmen Section of the Yellow River. *International Journal of Sediment Research* 23, 159-166.

Renner, J. 2012. Global irrigated area at record levels. Washington, D.C.: Worldwatch Institute.

Romm, J. 2011. Desertification: The next dust bowl. *Nature* 478, 450-451.

Sakthivadivel, R. 2007. The groundwater recharge movement in India. In: Giordano, M., Villholth, KG (eds.) The agricultural groundwater revolution: Opportunities and threats to development, CAB International, Wallingford, UK, pp. 195-210.

Selvaraj, K., Yoganandan, V., & Agoramoorthy G. 2016. India contemplates climate change concerns after floods ravaged the coastal city of Chennai. *Ocean and Coastal Management* 129, 10-14.

Shah, T. 2010. Taming the anarchy: Groundwater governance in South Asia. RFF Press, Washington, DC.

Shah, M., D. Banerji, P. S. Vijayshankar, and P. Ambasta, 1998. India's drylands.

Tribal societies and development through environmental regeneration. Delhi: Oxford University Press.

Sharma, N. D. 2007. Do you believe in a second green revolution? Current Science 92:1032-1033.

Shingi, P.M., & Asopa, V.N. (2002). Independent evaluation of check dams in Gujarat: Strategies and impacts. Indian Institute of Management, Ahmedabad, India.

Shiva, V. 1992. The violence of the green revolution: Ecological degradation and political conflict in Punjab. Delhi: Zed Press.

Singh, K., and V. Ballabh, 1996. Cooperative management of natural resources. Delhi: Sage.

Starke, L. 2008. State of the world: Innovations for a sustainable economy. New York: WW Norton and Company.

Singh, K., & Jagawat, H. 2011. The Sadguru Model of Sustainable Rural Development: An update. A report published by Sadguru foundation, Dahod.

Singh, R.K., Pandab, R.K., Satapathy, K.K., & Ngachana, S.V. (2011). Simulation of runoff and sediment yield from a hilly watershed in the eastern Himalaya, India using the WEPP model. *Journal of Hydrology* 405, 261-276.

Sinu, P.A., & Nagarajan, M. (2015). Human–wildlife conflict or coexistence: what do we want? *Current Science* 108, 1036-1038.

van Wiren-Lehr, S. 2001. Sustainability in agriculture: An evaluation of principal goal-oriented concepts to close the gap between theory and practice. *Agriculture Ecosystem and Environment* 84:115–129.

Varshney, A. 2000. Ethnic conflict and civic life: Hindus and Muslims in India. New Haven: Yale University Press.

Viju, B. (2013). Check dams to curb man-animal rift. 9/11/2016:timesofindia.indiatimes.com/city/thiruvananthapuram/check-dams.

World Commission on Dams (2000a). Large dams: India's experience. Earthscan, London.

World Commission on Dams. (2000b). Dams and development: A new framework for decision-making. Earthscan, London.

WWF, 2006. Living planet report. Gland, Switzerland: WWF International.

Xin, Q.C., Liu, L., & Shi, W.B. (2004). The design for the optimum height of warping dam. *Res Soil and Water Conservation* 11, 154-156.

Xu, X.Z., Zhang, H., & Zhang, O. (2004). Development of check-dam systems in gullies on the Loess Plateau, China. *Environmental Science & Policy* 7, 79-86.

Yuan, T., Li, F.M., & Liu, P.H. (2003). Economic analysis of rainwater Harvesting and irrigation methods, with an example from China. *Agricultural Water Management* 60, 217-226.

Zeng, Q.L., Yue, Z.Q., Yang, Z.F., & Zhang, X.J. (2009). A case study of long-term field performance of check-dams in mitigation of soil erosion in Jiangjia stream, China. *Environmental Geology* 58, 897-911.